Behaviour-Based Safety in Organizations

Saving Life before the Accident

Behaviour-Based Safety in Organizations

Saving Life before the Accident

Second Edition

H.L. Kaila

Professor and Head (Retd.)
Postgraduate Dept. of Psychology
SNDT Women's University
Mumbai
Emeritus Professor
JJT University, Rajasthan
&
BBS Project Director, GAIL India

TECHSAR™
NEW DELHI

Published and Printed by

Techsar Pvt. Ltd.
4435-36/7, Ansari Road, Daryaganj,
New Delhi 110 002 (India)
E-mail: tech_info@techsarworld.com
Website: www.techsarworld.com

ISBN 978-93-85909-45-0

Reprint 2026

Foreword

The National Policy on Safety, Health and Environment at Workplace declared by the Government of India aims at reduction of occupational injuries and diseases. The establishment and maintenance of a preventative safety culture is the ultimate goal of the policy. As 80% of accidents at the workplaces are caused by unsafe behaviours of employees, Behaviour-Based Safety Management (BBSM) assumes greater significance. Behaviours of the working people on the job are indicators of probable injuries and safety performance. The BBSM is a proactive and practical approach which genuinely influences the working people and helps them to follow safe work practices and develop safe habits on the jobs. This technique serves as a tool to correct their unsafe behaviour and achieve desired safe practices at the workplaces.

India is looking forward to a high growth in the manufacturing, construction and service sectors. The safety and health cannot but be part of this development and growth. Therefore, it is vey appropriate that at this crucial juncture, this book focuses on the emerging areas such as behavioural based safety process, BBS implementation including role of Steering Committee and training team, success stories, experiences of organizations which have implemented this method, etc.

This book has been authored by Dr HL Kaila, who is a Professor of Psychology in Mumbai. He is an industrial psychologist and is a regular BBS trainer in India. Dr Kaila has vast professional experience and has to his credit several projects with ICSSR, UGC, NSC, WHO, etc. He has conducted more than 300 BBS workshops covering over 12000 workforce for leading companies in India. He is sincerely committed to improvement of safety and health standards, professional knowledge and promotion of safety awareness among the working people and self-compliance by managements and all others concerned. I appreciate the laudable efforts put in by Dr. Kaila in compiling such a knowledge bank in the field of BBS.

I am confident that this valuable contribution by Dr Kaila will serve as an important source for the workforce, learners and academicians and reference for the professionals in the field of BBSM.

I take this opportunity to wish the book and its readers a great success.

Vinod Sant

VB Sant
Director General
National Safety Council of India

Preface to the Second Edition

Most employers and top managements have a faulty perception that safety is majorly a cost to the company. This perception is a constraint to consider safety as a value to save lives at the work-place. Every day, 6300 people die as a result of occupational accidents or work-related diseases, i.e., more than 2.3 million deaths happen per year (International Labour Organization, 2013).

Unsafe behaviours are at the core of all near-misses, injuries, and fatalities. If we control unsafe behaviours, we would not even have near-misses. The Behaviour-Based Safety (BBS) process helps maintain an account of human behaviour in terms of percentages of unsafe and safe behaviours of workers on a monthly basis.

BBS is a whistleblower on the behavioural risks which are apparently business risks like violating standard procedures and so on. BBS means when you are walking through your workplace and observing that somebody is working under such a risk, you stop for a couple of minutes and alert him for spot correction as a big brother and not as boss, and this correction procedure is developed in the organization as a regular way of work life involving everyone from top to bottom, and this entire BBS activity is measured in a scientific way each month.

A single unsafe behaviour can cause accident. Hence, the entire BBS implementation evolves around adopting a correctional procedure, not fault finding. Also, it's important to understand that culture drives behaviour, so you can't punish an individual.

BBS application is truly democratization of correctional power for developing a decision-making attitude to draw lines between what is safe and unsafe behaviour across work areas for a clear focus on core values, i.e., saving human life at every cost by everyone as a big brother, not as a big boss. If you hope to permanently put a stop to human, financial and production losses due to accidents at your work areas, then you need to essentially empower your manpower to get rid of their at-risk behaviours through BBS implementation and following it continuously by involving all. BBS is a significant contributing tool towards corporate safety management performance.

H.L. Kaila

Acknowledgements

I acknowledge the contributions of several organizations and their managements directly and indirectly in the formation of this manuscript.

1. Amara-Raja
2. American Society of Safety Engineers
3. Ampacet
4. Bajaj Auto
5. Bayer CropScience
6. BHEL
7. BPCL
8. Central Labour Institut
9. Chambal Fertilizers
10. Colourtex
11. Centre for Technical and Engineering Applications
12. ESSAR
13. GAIL India
14. Greentech Industries
15. HCC
16. Hindalco
17. HPCL
18. Indian Chemical Council
19. IOCL
20. ITC
21. Kalpatru
22. L&T
23. Mahindra & Mahindra
24. National Safety Council
25. Nuclear Power Corporation
26. Oil India Ltd
27. ONGC
28. RCF
29. Reliance Energy
30. Reliance Industries

31. SAIL
32. Sandoz
33. Serum
34. SMC
35. SNDT Women's University
36. Suzlon
37. Tata Institute of Social Sciences
38. Tata Motors
39. TUV
40. Ultratech
41. Vedanta

Contents

1

Behaviour-Based Safety (BBS) Concepts

1.1 INTRODUCTION

Behaviour-Based Safety (BBS) is a practice that ensures safe workplaces for employees in an organization, across all industries. As compared to other workplaces safety practices, BBS is the safety culture of the new-generation of managers. It is, simply put, a sort of employers' insurance against accidents at the workplace. BBS's first beneficiary is your safety department which acquires BBS-trained observers when it decides to implement BBS.

The main safety objective of every organization is to bring down the number of work-related accidents to zero. Since safety in the organization cannot be anything less than 100 per cent; the participation of each and every employee also has to be no less than 100 per cent. That is what Behaviour Safety is all about.

These objectives can be achieved only when we educate and train every employee on the concept and processes of behavioural safety. Before training ground-level employees, we need to sensitize and promote the concept BBS amongst the heads of departments and frontline managers.

"Before BBS implementation, safety procedures were like police inspections. BBS training has improved communication between safety officers and other employees, and also the attitude of employees towards safety. They are getting involved in BBS. The internal trainers of our BBS program have worked very well, as they know the job processes..."—the trained team members of an organization.

Research literature and industrial safety experts say that:

1. Almost 100 per cent of workplace-related accidents are caused by unsafe human acts and behaviours.

2. 30 per cent of the unsafe behaviours are noticeable at any plant at any given point in time.
3. 25-30 per cent of safety awareness is lacking among employees which gets reflected in their unsafe behaviours.

4. Unsafe behaviours are at the core of almost all near-misses, injuries, and accidental deaths. If we can control unsafe behaviours at our workplace, we may not have even near-misses.

BBS has also become successful because it ensures that employees are empowered and made responsible for their own safety, without supervision except at the basic level. BBS turns organizational safety into teamwork; it is a partnership between employees and their management.

1.2 BBS IS AN OHSAS-COMPLIANT PRACTICE

OHSAS 18001:2007 document is included in occupational health and safety **(OH&S)** management systems.

In this document, section 4.4.3.2 mandates the participation of, and consultation with, employees and also says that an organization shall establish, implement and maintain procedure(s) for **the participation of workers by their appropriate involvement in hazard identification, risk assessments and determination of controls.** "*Until we involve people, we cannot make our worksites safer,*" says a project manager at HCC.

OHSAS 18001:2007, clause 4.3.1 (*Planning hazard identification, risk assessment, and determining controls*) says that an organization shall establish, implement, and maintain procedure(s) for hazard identification and risk assessment that take into account human behaviour, capabilities, and also other human factors.

OHSAS 18001:2007, clause 4.4.2 (Competence, training and awareness) says that an organization shall establish, implement and maintain procedure(s) to make employees aware of OH&S consequences, actual or potential, of their work activities, their safety behaviour, and benefits of improved personal performance.

BBS processes include OHSAS 18001:2007 compliances on human factors in its training applications and helps organizations implement OHSAS 18001:2007 clauses mentioned above in this section. A BBS approach enables the participation and involvement of workers in hazard identification, risk assessments and determination of controls; and also makes workers aware of their unsafe as well as safe behaviours. BBS helps maintain an account of both safe and unsafe human behaviours in terms of measurable metrics (as percentages, for example) on a monthly basis.

This document shall help facilitate the implementation of the above three clauses that are a part of OHSAS 18001:2007.

1.3 THE BBS PYRAMID

The practice of BBS tackles behavioural problems leading to workplace mishaps at the root (see Fig. 1.1) Unsafe behaviours are at the core of almost every near-misses, injury, and fatalities at workplaces. If we can control unsafe behaviours, even near-misses can be avoided.

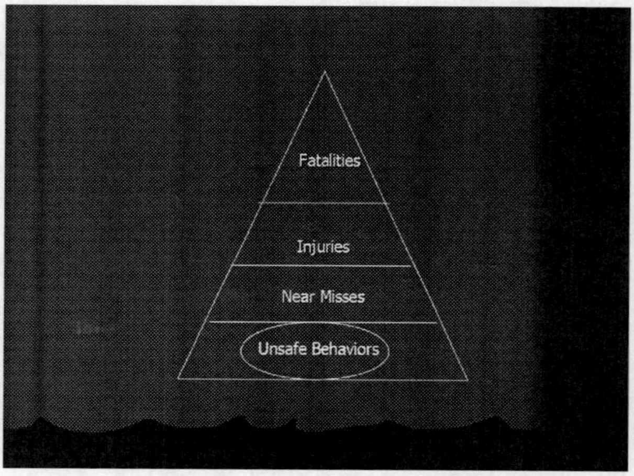

Fig. 1.1 The BBS pyramid.

BBS emphasizes that employees need to take ownership and responsibility of their safe as well as unsafe behaviours. If they behave in an unsafe manner, they are not to be punished but educated; and when they behave safe, they are to be encouraged. Both unsafe and unsafe behaviours are counted and displayed on the workfloor. Along with this, BBS also studies unsafe work conditions that lead to unsafe work behaviours by employees.

1.3.1 The Actual Impact of BBS Training

BBS is one of the latest approaches to safety at workplaces and is also one of the best. It does not involve micromanagement at any point, instead emphasis is laid on repeated education. BBS involves repeatedly going to an employee and making random observations till he or she starts practising safe work behaviours and learns the concept of self-observation and observing others for safe performance. Nine aspects that behavioural safety takes care of in order to prevent unsafe behaviours and promote safe behaviours are (see Fig. 1.2):

1. Personal Protective Equipment (PPE).
2. Housekeeping.
3. Using proper tools and equipment.
4. Body positioning and protection when at work.
5. Material handling.
6. Communication.
7. Proper following of procedures.
8. Visual focusing.
9. Use of mobile phones when at work.

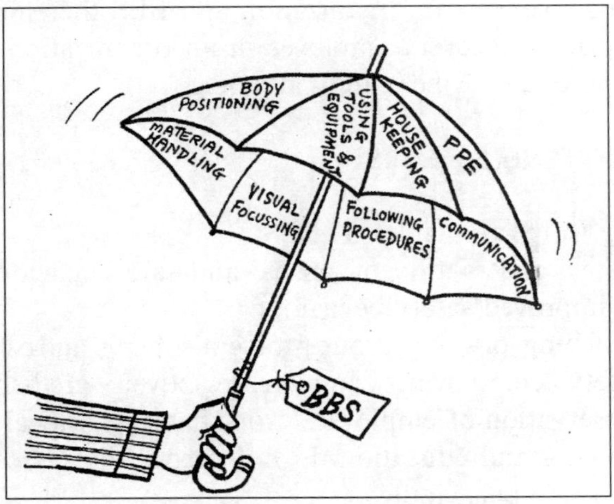

Fig. 1.2 BBS umbrella of protection.

With the help of this checklist, BBS-trained observers collect data on these nine parameters. Each observer provides feedback to at least one of his coworkers on a daily basis. This makes safety a daily practice and helps build a culture of safety. We can do a daily check on the increase or decrease in unsafe and safe behaviours by creating this database.

The decision making process in BBS is scientific and purely data-driven. According to this science, safety practices and behaviours have to be measurable in order to be effective and each employee can make a big difference in organizational safety. Employees are the basic source of expertise of behavioural change (observe and correct.) BBS begins by briefing sessions for all work areas and in all departments. BBS is a company-wide and employee-driven team effort, and enforcement, forced procedures, and micromanagement down to the bare minimum required. Its purpose is to identify safe and at-risk behaviours, identify possibilities that may lead to accidents and injuries, communicate these risks to employees, and then identify solutions. An implementation team (or BBS steering committee) monitors this entire cycle.

Essentially, BBS is not a management driven tool for safety but is an employee driven approach that is supported by the organization management.

Employee safety is too critical an issue to be left to just the safety department in an organization and BBS therefore makes sure that each and every employee in an organization is made responsible for his or her safety at work.

1.4 BASIC TENETS OF BBS

According to the principles of BBS:

1. Active observation of unsafe and safe behaviours leads to improved safety behaviour.
2. Listening, praising, group problem-solving, and celebrating safety achievements can increase actively safe behaviours.
3. Observation of employees' work habits is critical because training and education do not necessarily get converted into safe behaviour.
4. Attitude is internal, refers to thinking and realization; whereas behaviour is external, observable, and an active experience.
5. Both individual and social dynamics for safety need to be addressed.
6. It does not replace process-hazard evaluation, incident analysis, or environmental and cultural solutions. When at-risk behaviours are identified, comments are made on the organizational cultural factors that reduce or prevent such behaviours.

7. Effects of BBS-implementation can be seen by measuring an organization's safety climate and awareness before and after its implementation for understanding changes in safety performance.
8. Rules laid down in BBS can be customized according to the needs of a particular organization.
9. The following cycle is observed in all BBS processes: *Define* → *Observe* → *Intervene* → *Test*.

Fig. 1.3 BBS cycle in a nutshell.

10. BBS has shown positive results in terms of safe behaviour and reduction in the number of accidents across industries and countries.
11. BBS increases safe behaviours and therefore reduces injuries, illnesses and the related financial costs on both the organization and its employees.

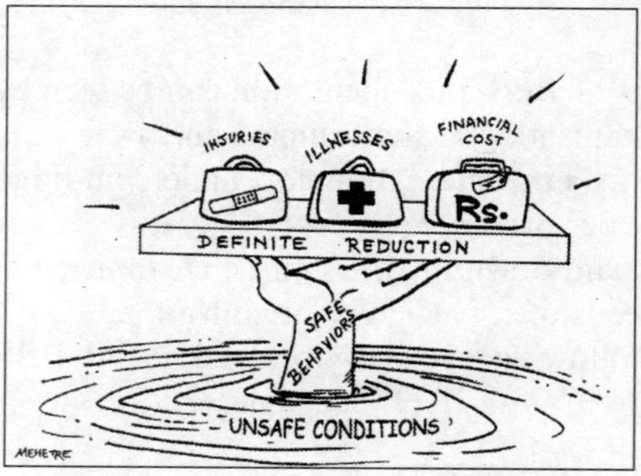

Fig. 1.4 How BBS helps organizations.

12. BBS processes comprise observation and feedback; along with a system of collection, analysis, and dissemination of data and a proactive support of management.

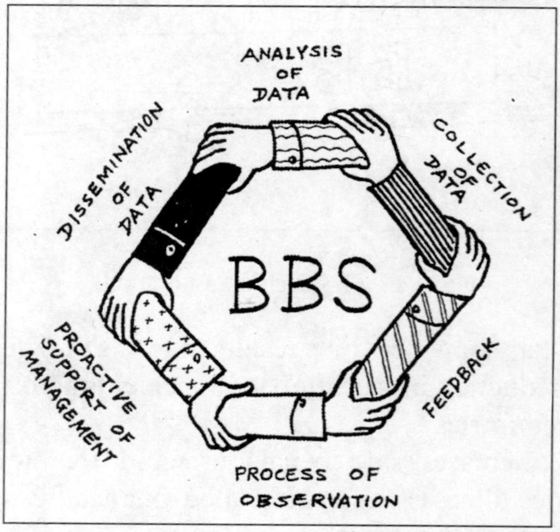

Fig. 1.5 How BBS works.

13. BBS is 'actively caring': it goes beyond the call of duty to safety for self and others.
14. BBS is a process with the purpose of preventing unsafe behaviours at workplace.

15. BBS involves peer-to-peer monitoring of work habits with the aim of reducing unsafe behaviour at a workplace.
16. BBS is more than safety regulations, but is a gradual process that involves educating employees.
17. BBS in no way involves disciplinary action, personal prejudice, or top-down implementation. It is based on praise, encouragement and reinforcement of safe behaviours.
18. BBS works bottom-up and not top-down like most other safety processes.
19. BBS is based on the established behavioural principles in psychology and work culture.

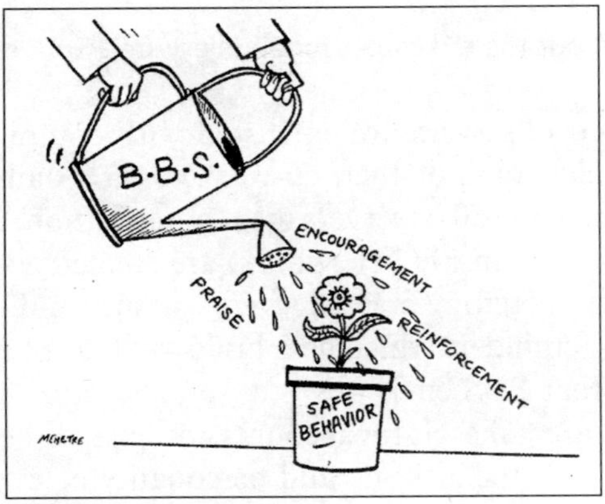

Fig. 1.6 BBS involves active participation of employees.

1.5 BBS AND ORGANIZATIONAL HEALTH AND SAFETY (OHS)

BBS is all about rectifying unsafe behaviours with the purpose of reducing the number of accidents and promoting safe behaviours in any organization.

Fig. 1.7 Roll out the red carpet for an injury-free work environment.

BBS-trained observers are entrusted with compiling data on the critical behaviours of their co-workers on a daily basis. For example, if there are 50 observers in an organization, 50 checklists (on safe as well as unsafe behaviours) are created everyday, 1500 checklists are therefore created every month and these then become daily reminders which help build a safety culture at work.

The ideal that BBS envisions is a Total Safety Culture (TSC) and this culture takes into account not just safety rules and practices but also the attitude and personality of each employee taking part in the process. This process, which helps create TSC in organizations is known as Behaviour Observation and Feedback Process (BOFP.)

A well-planned and implemented behavioural safety system should lead to: fewer accidents, mishaps, near-misses, and lower property damage; improved levels of quantifiable safety behaviours; reduced accident costs; increased reporting of defects, accidents; regular and rapid follow-up in case of any mishap; *and,* sharper and varied skills that enable positive reinforcement of safe behaviours.

To rephrase an often used line, BBS is by the employees, of the employees, and for the employees.

Fig. 1.8 BBS is practised by employees for their own benefit.

CHECK YOUR UNDERSTANDING (CHAPTER 1):

Instructions: *For each of the questions below, discuss with your learning partners and then make individual notes in your diary. You may refer to this manual while doing so.*

1. How can established but unsafe work behaviours be effectively checked?
2. Which is the most efficient way to implement BBS in your organization?
3. In your opinion, are no accidents or safety-related mishaps, and 100 per cent safety compliance achievable objectives? If Yes, how? And, if No, why not?
4. How can unsafe behaviours of employees be identified? And how can these be then converted to safe behaviours?
5. Why do you think employees do not practise safe behaviour at work even though they are aware of it?
6. Want to know the concept, the process of BBS and improve safety culture?
7. At what level in your organization should BBS implementation start? Give reasons for your answer.
8. Do you think a majority of organizations are aware of BBS and are ready to spend money and use company assets on its implementation?
9. How to involve and convince trade union leaders for BBS?
10. Discuss the following observation by the author which has come after years of being a top-notch professional in this domain: *It is said that safety flows from top management to bottom management, particularly in government organization. Top management approach is very casual towards safety. So it requires first of all changing the behaviour of top management instead of workers' behaviour or approach toward safety.*
11. Is BBS is similar to DuPont's Safety Training Observation Program (STOP)?
12. Approximately how much time would it take to make your organization BBS-enabled?
13. Make a list of organizations in India that have implemented BBS. How far has this implementation been successful?
14. In the BBS feedback process, who gives feedback and should it be given on the spot or later?
15. How can we effectively change work behaviour of employees?

16. What are the steps in designing and writing of BBS-related content for training: (i) senior management; (ii) trainers; and (iii) learners.
17. What is accident proneness?
18. Observing behaviour is alright but then, how can we influence our peers for changing their safety behaviours? Do we have a scientific process for this?
19. Mostly accidents happen due to ignorance or over-confidence and also to other reasons when most of us don't wear helmets on a two-wheeler or seat belt in a four- wheeler. How are we confident that we can influence our associates towards desired behaviour patterns?
20. What are the steps/ processes of BBS implementation?
21. How is feedback given and corrective action taken when implementing BBS?
22. How can safety awareness be instilled at subconscious levels amongst employees?
23. How can employees that come to work under the influence of alcohol or drugs be dealt with under BBS?

2

Behaviour-Based Safety (BBS) Processes

The process of BBS comprises repeatedly going to an employee and making random observations till he reaches safe behaviours and learns the concept of self-observation and observing others for safe working behaviours.

2.1 FACTORS THAT ENSURE SUCCESSFUL BBS IMPLEMENTATION

Listed below are the various processes that are involved in BBS implementation. These processes are described in details later in this chapter.

- Observation and ffeedback.
- What does a BBS observer gain by actively caring?
- Problems encountered in observing and giving feedback.
- Reasons BBS observers may fail in observation and submitting feedback properly.
- BBS observers require good interpersonal skills.
- BBS observers' test of excellence.
- The eight P's BBS observers should follow.
- The concept of an active observer.
- What excited the observers.

2.1.1. Observation and Feedback

Observation and feedback by BBS observers involves the following:

1. Collecting observation data on specific safe and unsafe behaviours and classifying them by department, date, month, and time.
2. Regularly monitoring safety behaviours.

3. Behavioural change precedes attitudinal change. We see tiger run and then experience emotion of fear. Realization occurs better by doing.
4. BBS is basically a Behaviour Observation and Feedback Process (BOFP).
5. Behavioural change brings 'attitudinal change', but the reverse might not always hold true. BBS observers should therefore focus on behavioural changes and attitudinal changes will logically follow.
6. Critical behaviours can be listed in checklist based on earlier accident and injury records, and also by brainstorming.
7. Different observers will notice different safe and unsafe behaviours, which is why employees need to observe each other.
8. Observers should chalk out observation routines and continuously keep improving safety processes.
9. Feedback can be on-the-spot, or weekly or monthly. Feedback can also be given in a graphical format by using infographics.
10. Feedback is an interaction based on genuine concern; we should keep an open mind when giving and receiving feedback.
11. Feedback should be given one on one. Safe behaviour should be reinforced and appreciated to set examples for others to follow.
12. Put in Hindi, BBS is all about *dekho* (observe) and *bolo* (*give feedback*) with sensitivity and concern.
13. Employee, name should not be recorded in BBS checklists and feedbacks.
14. Observers should concentrate on safe and unsafe behaviours of coworkers.
15. Observation and coaching might take some time to be accepted by coworkers.
16. Observation and feedback skills improve with practice and also by using a standard checklist.
17. Observations should be random (not scheduled) throughout the week.

18. Observer's feedback should be constructive. For example, instead of just saying wrong tools were used to do a job, it should also be mentioned which is the right tool that should have been used.

Fig. 2.1 The two tenets of BBS as said in Hindi.

19. Observe both safe and at-risk behaviours and record detailed comments for problem solving, follow up, etc.
20. Observe in two-member team, in which one should be experienced observer and the other a trainee observer.
21. Observers should provide feedback on safe and unsafe behaviours to an observee in the form of a checklist.
22. Observers should immediate feedback for correction of behaviours.
23. Safety coaching fosters open communication about safety, and serves as a constant reminder for workplace safety.
24. Safety observers and coaches in every department should use different critical behavioural patterns (e.g., use of PPE, body positioning, use of tools, etc.) on their respective checklists and coaching processes (e.g., problem solving, follow-up).

25. Some characteristics of a good BBS observer are:
    - Concern for others.
    - Self-initiation for correcting the observee.
    - Ability to develop a mutually agreeable insight on safe behaviour.
    - Training the observee in self-observation.
26. In order to make BOFP easier, the observer should know how to communicate in languages that employees understand well, get to the intellectual level of employees, and should never let his ego harm his job as an observer and safety coach. One may be a manager as BBS observer but during BOFP he is not a manager; and he is playing the role of BBS observer so that BOFP functions as if it's peer-to-peer.
27. Observers need to focus on behaviours rather creating issues out of unsafe work conditions or something else that is the management's responsibility. Also, observers should not expect work conditions and behaviours to change overnight.
28. Observers should keep observation and feedback process on daily basis.

2.1.2 What a BBS Observer Gains by Actively Caring

Actively caring (listening, praising, accepting) are very important aspects of BBS, and one needs to understand what does a BBS observer gain from actively caring.

Here are some responses of the BBS training participants:

- Positive and safe work culture.
- Increase in production.
- Increase in job satisfaction.
- Camaraderie amongst employees.
- Ownership of safe work processes.
- Enhancement of team spirit.
- Confidence amongst employees.
- Employees stop shying away from voicing their opinions and ideas.
- General happiness and peace of mind amongst employees;
- Free and frank exchange of ideas.
- Reduction in usage of unsafe work practices.

Fig. 2.2 Steps a BBS observer should take.

All the above responses clearly indicate that BBS helps in building positive safe work cultures in organizations. Individual observers can add creativity to behaviour categories of BBS checklist.

2.1.3 Problems Encountered in Observing and Giving Feedback

Following are a few factors that can be a hurdle in proper observation and therefore improper feedback later:

1. Observer handing over checklist to observee and asking him to fill it.
2. Only observing and not giving feedback.
3. Observer and observee not agreeing with each other.
4. Observer not allowing observee to speak.
5. Comments column not filled.
6. Feedback comprising only criticism without any appreciation.
7. Inadequate time for BOFP.
8. Observee not accepting observer's feedback.
9. Observee might over-react if given negative feedback.
10. The observer might think whom to observe.

2.1.4 Reasons BBS Observers may Fail in Observation and Submitting Feedback Properly

A few more problems that are encountered in observing and giving feedback are:

- Observee responding in an aggressive manner.
- Observer giving feedback in the presence of other employees.
- Observer not following Standard Operating Procedure (SOP); He himself not following SOP.
- Communication gap between observer and observee.
- Feedback not being given in a proper way.
- Lack of rapport between observer and observee.
- Observer not knowing observee's job well.
- Observer who is not very patient when interacting with observees.
- Observer losing objectivity and getting personal with the observee.
- Lack of proper follow-up once feedback is given.
- Observer not interacting on a one-to-one level, but acting superior and more knowledgeable.
- Observer unable to convey the importance of an injury-free work culture.
- Observer cutting corners, for example, recording the behaviours of more than one employee on the same checklist.
- Observer not being proficient in giving feedback.

While making sure that these points do not hamper his job, a BBS observer should also not make the observation and feedback processes too much formal; or be dominating in his interactions with the observee. An observer should be sensitized to the fact that observees should be treated on a peer-to-peer level. A feedback based on mutual discussion and respect between observer and observee on safe as well as at-risk behaviours helps create a safety culture much faster.

Anyone, a manager or a worker, can be a BBS observer once he or she has been properly and formally trained. In an Indian chemical company, for instance, it was seen that three groups of BBS trainees, mostly managerial employees, successfully conducted BOFP on groups of contract workers in three different geographical locations. This makes it evident that BBS-BOFP brings managers and workers closer and enhance camaraderie. It makes workers feel that senior people in the organization care for them for their safety and well-being.

2.1.5 BBS Observers Require Good Interpersonal Skills

The following interpersonal skills are critically required when any observer is giving his or her feedback:

- Trust-building ability; not displaying seniority, pride, or ego.
- Having high levels of patience when giving feedback; not being too formal or serious, but giving feedback in a formal, understandable, and may be even humorous manner; ability to listen and understand the observee speaks; displaying empathy with the observe; always being the more active participant in the feedback process and never looking low on energy; becoming a friend and not a boss; not to be under any kind of personal or professional stress when giving feedback; *and*, appreciating the safe behaviour of employees along with telling him where he has to do better.

The above interactive skills, when practised over a period of time, help a BBS observer in self-development so that the he or she can conduct BOFP effectively through active caring. These observers are not police inspectors going around and looking at unsafe conditions but they observe safety behaviours of employees/coworkers on a daily basis.

Fig. 2.3 How a good BBS observer should be

2.1.6 BBS Observers' Test of Excellence

For purposes of self-appraisal, BBS observers should grade themselves on a scale of 0-10 on the following parameters:

1.	Quality of observation.
2.	Satisfaction with BOFPs conducted.
3.	Understanding and filling of the checklist.
4.	Quality of interaction with observee.
5.	Comfort level of communication skills.
6.	Computer skills.
7.	Interpersonal skills (with reference to giving feedback).
8.	Interest in safety.
9.	Level of understanding of BBS.
10.	Willingness for development and growth as an observer.

2.1.7 The Eight P's BBS Observers Should Follow

Risk based conversations between an observer and observee when feedback is given is critical for safety in the organization. Making sure that there are no communication gaps when it happens can be ensured by following these eight Ps.

1. ***Parikarma*** (Tour): Tour your entire unit/plant when making observations.
2. ***Prashna*** (Query): Put safety related questions to observees.
3. ***Prashansa*** (Praise): Praise observees for safety procedures they are already following.
4. ***Parivartan*** (Change): Change his unsafe behaviours to safe ones.
5. ***Prashiksan*** (Training): Educate and re-train him in safe practices where required.
6. ***Pratigya*** (Oath): Make him swear under oath from that he will not repeat unsafe behaviours.
7. ***Prachar*** (Spread the word): Ask an observee to observe his colleagues.
8. ***Prarthana*** (Request): Repeatedly request observees to follow safety processes.

2.1.8 The Concept of an Active Observer

An *active observer* has the following characteristics:

1. Not critical and shows proactive behaviour.
2. Rectifies unsafe behaviour there and then.
3. Works with a positive and persuasive attitude.
4. Appreciates safe behaviour.
5. Takes a gentle and patient approach.
6. Action and solution oriented.
7. Explores unconventional ways and means.
8. Empowers colleagues and follows the bottom-up approach.
9. Is in continuous contact with other observers.
10. Result and analysis oriented.
11. Never ignores unsafe behaviour, but always stops and corrects.
12. Rectifies unsafe behaviour on the spot.

2.1.9 What Excited Observers Say

1. Earlier I thought of BBS as my optional work, now it has become a part of my life.
2. Save life is a great mission.
   BBS is our holy ambition.
3. Unsafe behaviour and unsafe conditions.
   Neither will we make it happen.
   Nor we will let it happen.
4. The most important aspect of BBS is the 'number of observers in the plant' and 'how actively they act in correcting unsafe behaviours on regular bases'.
5. It's better to wear a little sub-standard helmet than not wearing one at all.

3

Implementation of BBS

Now that we know how important BBS is to an organization in terms of both production, and efficiency and well-being of employees, we shall see how to go about its implementation.

3.1 THE ROADMAP

BBS is all about involving people across departments in an organization with a bottom-up approach. In order to implement BBS at workplace and create a culture of safety, following are the steps that should be followed:

1. BBS-awareness programmes for management staff of all levels.
2. One-day awareness training of employees across the plant.
3. Then, select observers from BBS trained employees, 20 per cent of all employees should be made observers, and form a steering committee with 8-10 people from BBS trained employees in it. An organization may select eight observers from every department which will make sure that there is one person to observe each behavior category.
4. One-day in-plant practical training of BBS observers and steering committee members on how to set up the observation process, how to develop the measure, how to making accurate and consistent checks, job details of the steering committee, etc.
5. Publicizing the BBS-implementation project through banners and posters across the plant.
6. Ensure compliance to safety standards and provisions of PPE and safety resources.
7. Regular involvement and engagement of managers with the BBS project.
8. Monthly meetings of observers and quarterly external reviews.

9. At this stage BBS observers and steering committee members are fully prepared to implement BBS at workplace. The above exercises have been successfully tried in India also.

3.1.1 BBS Training Project and Pre-implementation Activities

BBS training must be inaugurated and concluded by the plant head or senior-most person to lead the project.

1. A certificate of attendance must be issued to all BBS training participants for motivation.
2. BBS training must be evaluated with proper feedback at the end of the workshop.
3. A sapling must be planted at the end of the programme to mark the launch of the BBS project.
4. Group photograph of BBS participants must be used along with BBS message for company's newsletter or intranet.
5. BBS banners and posters need to be displayed across work areas and also on the organization's entry gates under the provisions of OHSAS 18001.
6. Stickers with "BBS Observers" printed on them should be pasted on the helmets of all BBS observers.

The success of any BBS programme in an organization depends on the following three factors:

1. Management's continuous monitoring of BBS's progress and its activities.
2. Frequency of observations.
3. Regular meetings of BBS observers with the steering team.

Checklist for assessing month-wise functions of BBS project leaders/plant/line managers:

	Functions	Week 1	Week 2	Week 3	Week 4
1.	No. of plant rounds made				
2.	No. of observers met and interacted with				

3.	No. of unsafe behaviours observed and closed				
4.	No. of unsafe conditions observed and closed				

(**Note:** *it's very important for plant/line managers to attend BBS steering team meetings every month*).

3.2 BBS IMPLEMENTATION AND STEERING COMMITTEE

1. Problems that are faced during implementation of BBS include: lack of workforce; unsafe behaviors not defined with precision; and, accident and near-miss injuries not analyzed properly for targeting accident causing behaviours.

Fig. 3.1 Problems in the implementation of BBS.

2. BBS may start small and build which means that it can begin in just one department. BBS is customizable and there is no strict and standard way to implement it.
3. The implementation in the plant must begin as soon as BBS training in order to make sure that the zeal of just-trained observers does not lose its momentum.
4. Observers have to work within their limitations without arguing with observees or stressing them. They should

also keep revising their patience threshold levels and concentrate on what percentage of change in safe/unsafe behaviors they achieve. They should keep revising their goals in order to achieve higher levels of safe behaviors from one month to the next.

5. People are asked to volunteer to either become observers or a part of the steering team. These people should carry out their responsibilities and duties with due diligence.
6. BBS needs to be integrated as an organizational system. It's a new initiative, new emphasis, and new direction on safety.
7. BBS's secret of success is that control is in hands of employees, which makes them feel empowered and responsible.
8. BBS should have pre- and post-measures of safety awareness levels of employees in all departments through safety awareness surveys for comparing its effects before and after implementation.
9. The speed of BBSsuccess depends upon existing injury and accident rates and readiness to implement it.
10. BBS's success depends on the efficiency of the steering team, which clearly defines its roles and responsibilities.
11. Behavioral safety management support and leadership of first line and senior managers can be measured quantitatively.
12. Implementation teams and the steering committee should routinely monitor the progress of BBS implementation.
13. In the long run, BBS is cost-effective as it reduces accidents, which would otherwise have cost the organization a lot of money.
14. Initially, BBS can be implemented in just one or two departments and then introduced in other departments.
15. Normally, self-nominations are invited from amongst the BBS trained employees to be observers, and they are further trained through a 2-day in-plant training.
16. It requires a steering committee comprising a senior manager, a frontline manager and around ten BBS trained safety observers.

17. Implementation of BBS can be achieved without hiring the services of a professional consultant.
18. BBS is an employees movement on behavioral safety with support and commitment of the management.
19. Management allows observers time to conduct observation tours, conduct feedback session, and organize data analysis to display, without which BBS will fail.
20. BBS is an active partnership between managers and shopfloor members of the steering committee.

Fig. 3.2 Management has a big role to play in successful implementation of BBS.

21. Managers in the steering committee should take active part in the process and not leave all the job to observers. Managers' role is critical and includes appreciating BBS observers, analyzing data, etc.
22. Managers and HoDs must appreciate observers' performances as they show progress in the percentage of safe behaviours.
23. Researches indicate that BBS has reduced accident rates by 40-75 per cent within six months to a year of implementation.
24. In India, ITC, Colourtex, HPCL, IOCL, GAIL, HCC, Pidilite, SAIL, Bajaj Auto, Reliance, Essar, Bayer

CropScience, Tata group, along with many others organizations, have implemented BBS.

To begin implementing BBS at the shop-floor, let the BBS trained employees complete a checklist of BOFP daily and continue imparting BBS training. More the number of BBS trained employees, more will be the number of BBS checklists coming in. Hence, you will find increasing data on unsafe and safe behaviors being collected on a daily basis.

A senior operator once said, "We can take percentage of safe behaviours in the plant up to 90 per cent very easily, managers must appreciate our performance, here we perform and *chamchas* (sycophants) get recognition."

BBS' approach is non-threatening and actually facilitates existing safety systems.

3.3 PROBLEMS BBS' IMPLEMENTATION MIGHT FACE

Responses of 55 BBS trainees when asked about problems in implementing BBS in an organization:

- Overstressed: the observer may feel that their workload has increased.
- What will I get? The observer may feel BBS' implementation is a fruitless attempt for him personally.
- Whether observee will take it in the right spirit?
- Fear of workers' buy-in to BBS.
- Taking advantage of BBS – not doing regular work.
- Removing existing unsafe conditions first.
- Integrating BBS with already existing committees in the organization.
- Manpower and man-hours that are spent in BBS implementation.
- The extensive paperwork involved.
- Interpersonal communication gaps.
- Initial launching problems.
- Discussing names of observees in BBS steering committee meetings.
- Calling observees in BBS committee meetings for discussing their unsafe behaviours.

While implementing BBS in an Indian organization, an employee said, "You are policing us; we are doing this job for 16 years now." According to a manager, "it took us six months to change the mindset of our shop-floor workers; we had a lot of difficulty in changing their behaviour, behavioral change is a very slow process, now people have started doing it."

3.4 PREFERRED ACTION PLAN FOR BBS IMPLEMENTATION

The following action plan is the most widely used for BBS implementation:

1. A commitment from the management.
2. Training of employees across all departments in the plant;
3. Selection and training of observers from amongst BBS trained employees.
4. Formation of a 8-10-member-strong steering committee from BBS trained employees.
5. Develop checklists;
6. Selection of areas and departments for behavior observation and feedback processes.
7. Develop a timetable for BBS implementation.
8. Documentation, analysis and display.
9. Review to determine progress of implementation in the organization.
10. Allocate time for BBS observations, meetings, etc.
11. Cut-off date for launching BBS.

3.5 RESPONSIBILITIES OF BBS STEERING COMMITTEE

The major function of the BBS steering team is to regularly mentor, monitor and motivate observers in their respective plants.

The 15 tasks this team is entrusted with are:

	Checklist points	Yes	No	Comments
1.	Is the observation checklist fine (or needs revision)?			
2.	Is observation by an observer once a day sufficient?			

3.	Is monthly progress in terms of percentage of safe/at-risk behaviors across plants satisfactory?			
4.	Are fortnightly or monthly meetings of BBS steering team/ observers being held regularly?			
5.	Are observers doing BOFP daily?			
6.	Is data being regularly entered by observers?			
7.	Is data analysis being done every tenth day across plants?			
8.	Are BBS progress charts being displayed across plants?			
9.	Are monthly progress reports being sent by the plant/ QHSE/HR Heads?			
10.	Are survey report actions being closed across plants?			
11.	Are issues of individual observers being addressed?			
12.	Is follow-up by the external reviewer happening regularly every month?			
13.	Is the BBS project getting appropriate propaganda/ advertisement through banners and posters?			
14.	Are the observers being recognized/ appreciated?			
15.	Is the top management leading, supporting, and reviewing the implementation programme?			

Normally, the steering team meets once or twice every month for review/discussion.

4

Organizational Cases on BBS in India

This chapter includes original organizational cases and field observations prepared during BBS training interventions over a period of twelve years from 1997 to 2012 across organizations (such as, petroleum, engineering, automobile, cement, power, chemical, pharmaceutical, etc.) as a part of an on-going national action research survey of BBS in India and included 1751 executives and 713 workers in 64 organizations. It is assumed that the information gathered from this longitudinal nature of the research and the robust sample size shall be considerably useful for human resource and safety professionals when they implement the concept and process of BBS for reducing of accidents and promoting safe behaviours for developing an injury-free culture in their organizations.

4.1 ANALYSIS OF UNSAFE AND SAFE BEHAVIOURS

One hundred and thirty seven employees from five organizations, as indicated in the *Table 4.1* below, were asked to report their safe as well as unsafe behaviours which they engage in on a day-to-day basis at home, workplace, and when on their way from home to work and from work to home. An analysis shows that an average of 52.6 per cent practise unsafe behaviours and an average of 47.4 per cent practise safe behaviours.

Table 4.1: Unsafe and safe behaviours employees themselves engage in on daily basis

Organization	Number of employees	Number & percentage of unsafe behaviours	Number & percentage of safe behaviours	Total Number of behaviours
1	30	311, 48%	330, 52%	643
2	28	230, 49%	239, 51%	469
3	23	158, 54%	135, 46%	291
4	31	219, 47%	243, 53%	462
5	25	226, 65%	124, 35%	350
Total	137	Average 52.6%	Average 47.4%	

This analysis shows that while every employee has engaged in safe behaviours, they have simultaneously engaged in an unsafe behaviour, more than 50 per cent of the times at home, workplace and on the road while going from residence to their office and back. Therefore they need someone else to look after their safety which is emphasized by the principle of BBS that people need to take care of safety for each other, i.e., peer-to-peer safety.

These employees from various organizations were further asked to report unsafe behaviours of their coworkers as perceived during the last one year. *Table 4.2* reveals that almost every worker has practised, on an average, 3.5 unsafe behaviours at their workplace on daily basis which can convert into an accident, near-miss or it can also be fatal depending upon the activity workers engage in.

Table 4.2: Unsafe behaviours of coworkers

Organization	Number of employees	Number of unsafe behaviours perceived	Average unsafe behaviours
1	21	106	5
2	21	61	3
3	17	104	6
4	16	81	5

5	19	83	4.5
6	26	105	4
7	27	81	3
8	19	68	4
9	34	113	3.3
10	30	97	3.2
11	30	79	2.6
12	30	59	2.0
13	26	70	2.7
14	24	66	2.8
15	22	38	1.7
			Grand Average 3.5

4.2. CASE OF A CHEMICAL FERTILIZERS COMPANY

Trained BBS observers in Rajasthan conducted behaviour observation feedback process (BOFP) and observed the number of unsafe/safe behaviours and unsafe conditions. Table 4.3 reveals that the number of safe behaviours is increasing everyday/shift, the number of unsafe behaviours is decreasing everyday/shift, and the number of unsafe conditions is also decreasing everyday/shift.

Table 4.3: Safe / unsafe behaviours and conditions

No of BBS observers	No. of safe beh. 1st Shift	No. of safe beh. 2nd Shift	No. of unsafe beh. 1st Shift	No. of unsafe beh. 2nd Shift	No. of unsafe conditions 1st Shift	No. of unsafe conditions 2nd Shift
29	107	103	59	33	41	31
30	141	155	49	27	47	23
33	217	235	47	29	03	01
36	243	274	37	14	01	01

4.3 CASE OF A POWER COMPANY

The 20 technicians on the first day and the 13 engineers on the second day in the state of Karnataka were trained in one-day workshop on BBS. They were asked to indicate the unsafe behaviours they and their coworkers regularly engage with, the results are as below:

	No. of Individual unsafe behaviours	No. of unsafe behaviours of coworkers
Technicians:	198/20 = average 10	74/20 = average 4
Engineers:	88/13 = average 7	61/12 = average 5

This shows that the unsafe behaviours are possessed by each individual and also their coworkers which is the root cause of any near-miss, injury or accident.

During the BBS training, the 20 technicians and the 13 engineers were also asked to observe their coworkers about their safe and unsafe behaviours at the workplace, the results are as below:

	1st shift		**2nd shift**	
	Safe beh.	unsafe beh.	safe beh.	unsafe beh.
1st day	105	73	142	38
2nd day	64	33	82	15

The data in the two tables above reflects that due to BBS training, the safe behaviours have gone up and unsafe behaviours have reduced from 1st shift to 2nd shift.

It was also observed that **almost 1 per cent risk reduction is contributed by each BBS trained observer**, which indicates that more the observers, more the number of observations, and more the percentage of risk-reduction in the organization (*Monsteller, 1989*).

- The technicians and engineers were asked as to what did they like best about BBS and the results are as follows:
- BBS induces self-introspection and alertness among workers.
- It's a team-building and increased interactional process.

- It is about human safety.
- One is not forced to change in BBS.
- It involves two-way communication.
- It is a technician-based approach.
- It is not about punishing unsafe behaviour.
- We can change attitude by behavioural change.
- It helps building a positive safety culture.
- It is a systematic feedback approach (Marsh, *et al.*, *1998*).

The technicians and engineers discussed that the observers need to discuss both creating safe environment and promoting safe behaviours in their monthly meetings. The observers need not do mind-reading of an observee during BOFP. Just keep doing BOFP in a routine manner to ascertain its cumulative effect in terms of percentage of safe and unsafe behaviours at the workplace. The observers also need to make a weekly observation schedule in advance so that they can cover BOFP across departments in their organization.

The technicians and engineers wished to know more about how to regularize BBS, different types of PPE, whether BBS is also useful in ones' life, how to achieve cooperation between observer and observee, how to reduce personal stress of an observer, whether BBS is really effective, whether to correct behaviour or attitude, how to draw an action plan to launch BBS, can BBS be implemented in a risky work environment, is BBS according to OHSAS, whether observers should remain same or change.

4.4 CASE OF A STEEL COMPANY

Two hundred and fifty senior management staff and 313 non-management staff of a large steel company over two days in the state of West Bengal in India considered the following aspects of BBS:

1. If management is not taking action for correcting unsafe conditions or not supplying proper PPE, does it mean that the organization is not ready for BBS? Is it not that involving employees under BBS project would boost or encourage management to be proactive and resolve the pending issues.

"Heads of Departments are the real bottlenecks in safety management that is why I have arranged BBS awareness training first for them"–GM safety.

2. The **middle management can no longer sit on old unsuccessful stories**. They need to build success stories in creating safety culture under BBS programme.
3. It's often quoted that 90 per cent or more of the accidents are due to unsafe human acts or behaviours. In reality, these unsafe behaviours on daily basis exist in our organizations, and we are sitting on thousands of unsafe behaviours at our workplaces. We need to recognize and find ways to correct them by way of BBS.
4. **Labor unions and workers in India are positive about getting involved in BBS**. They have also expressed a strong concern for imparting BBS training to contract workers also, as almost 50 per cent of the workforce in the organizational premises is contract workers. "*Best safety device is a careful worker which we can achieve through BBS*", said a general manager.
5. Some departments/jobs have more at-risk behaviours than others depending on criticality of the nature of work. These jobs deserve an early BBS attention.
6. Other concerns expressed are as follows:
    - OHSAS is incomplete without BBS and without involvement of workers, OHSAS is helpless unless and until people actually participate.
    - BBS is a joint venture on safety between management and non-management employees. BBS is a confidence building measure on safety.
    - Zero accident does not mean that we are safe.
    - Certification from international bodies should not lead to complacency.
    - Behaviour modification in BBS is a challenge.

4.5 CASE OF A CEMENT COMPANY

Three days of interaction with engineers in BBS in the state of Maharashtra brought out the following issues:

1. Three groups of 88 engineers observed 239 unsafe behaviours of the workers during three days of training on BBS. On an average, three unsafe behaviours per worker were identified which means a total workforce of 710 shop-floor employees possess 2030 unsafe behaviours which is a serious concern safety for an organization. Similarly, a BBS survey can indicate thousands of unsafe behaviours existing in an organization. A manager said, *"Though we are aware of the magnitude of unsafe behaviours in our organization we did not know how to control them."*
2. The engineers of the company realized that feedback makes a difference in safe behaviour, not writing name on the BBS checklist makes a person open, recording on the checklist boosts interest of an observee, and the focus is better on safety with BBS approach. They also felt that the department-specific checklists can be prepared as behaviours are job-specific.
3. BBS observers opined that the BBS core team of 10-12 members should comprise departments of the organization and they would also simultaneously make a list of unfinished/pending unsafe conditions for regular follow-up with management. *"Safety should be there in the behaviour of human beings which is lacking,"* The unit head said.
4. **An observer is a role-model** for his observees for safe behaviours to follow. In high-risk areas, the observers minimize the risk and increase alertness on part of observees. Observers' names should be displayed in their respective department. It is not only observers who run the show; observees also take care of each other's safety.
5. In BBS, workers are involved as well as accountable for safety in the organization. A manager said, *"BBS is the life-line and passport of safety"*.
6. About 50 per cent of unsafe behaviours are known to people at the workplace, yet they are not pointing out. It is non-performance of safety behaviour. When unsafe behaviours are existing in an organization for a long time and safety has not been enforced or sustained it is **unsafe social behaviour**. In BBS we create safe social behaviours

where all workers join together and involve in creating safe culture. "*For 20 years of my career I have not thought of safety, safety must have begun when we started our career,*" a manager said.

7. What should be the **plan of action after having implemented BBS**? Basically, three things: estimating scope of re-training for observers, assessing role of BBS steering committee, and determining progress of BBS month-on-month basis.
8. **BBS provides additional safety force** in the organization. Training to all employees maintains uniformity of BBS values and culture. So training at all levels is essential either in small or larger groups. *"BBS if implemented is excellent for organization, if not, then also it is good, training in BBS provides a different mindset towards safety of people"*, an HR manager said. *"BBS would spread like a mobile"*, a trained BBS worker said.

4.6 CASE OF A CHEMICAL MANUFACTURING COMPANY

In the state of Gujarat, 53 senior and middle management employees (who had work experience between 10 to 30 years) participated in a one-day BBS workshop and brought out the following issues:

1. On an average 2.5 unsafe behaviours were observed at the shopfloor which means about 1125 unsafe behaviours existed in the organization at present (450 employees × 2.5 unsafe behaviours = 1125).
2. There are two kinds of employees as far as safety consciousness is concerned. One kind of employees who have internal locus of control for safety meaning they are internally conscious. The other kind of employees who are externally conscious meaning they require external stimulus to alert them regularly.
3. How to incude contract workers under BBS when a production department has 50 permanent employees and 250 are contract workers. This can be done in a phased manner, first train regular employees, and in the second phase train contract workers. Both need to be oriented for BBS concepts.

4. Whether fear factor is necessary for creating safety culture. No, because fear or punishment will not give sustainable result in changing unsafe to safe behaviours.
5. Should BBS be a man-specific or activity specific? It is of course activity or behaviour specific. It is not a name or blame game.
6. Should observers do BOFP in their respective departments or across departments? Initially, restrict to respective departments, later on when observers mature in BOFP, they can be assigned across departments.
7. How much time BBS would take to give results, what changes can be acquired through BBS. Surprisingly, BBS starts giving results quickly. More the observers, more the observations, more the safe behaviours. The outcomes or changes are the reduced unsafe behaviours, safe working conditions, safe culture, etc.
8. Is it a new approach which would be out soon like QC? Not really, because BBS is a data-driven approach (Pearse, 1997). What gets measured gets done. As long as it gives results in terms of reduced unsafe behaviours, it is most likely that BBS would stay in the organization. It prevents accidents and accident-related costs.
9. Is it Gandhi-giri (soft approach)? BBS is a soft but its data-driven approach.
10. The participants had mutually set the starting date and closing date for launching BBS, held FPR (first person responsible) for BBS, and set a date for microplanning details to be ready within a week's time.
12. Hence, this one-day BBS workshop was considered to be a pre-launch of BBS which provided a clear **roadmap for launching BBS.**
13. Many times observees are not aware that they are engaged with unsafe behaviours, that is why, the observers have to be very alert in their observations and save their coworkers from these unsafe behaviours which can be very harmful.

4.7 CASE OF A PHARMACEUTICAL ORGANIZATION IN GUJARAT

One hundred and eighty six employees including workers and executives were trained on BBS. The main observations that came out of the discussions in six days of training are as follows:

1. Observers noticed 3.5 unsafe behaviours on an average per worker being practised at their workplace on daily basis.
2. Observers found 34 unsafe conditions in observation tour and corrected 30 on the spot.
3. An observer first observes and thereafter provides feedback to an observe.
4. An observer may observe 3-4 observees at a time while filling the checklist for 9 critical behaviours.
5. Observees educate each other if an observer gives feedback effectively.
6. Normally, 10 minutes are sufficient for making observations and feedback.
7. While giving feedback to contract workers, it is important to give feedback simultaneously to their supervisor about the at-risk behaviours observed.
8. Observers can observe anybody (same cadre or above) and give feedback for correction.
9. Though 20 per cent of observers are selected per department in an organization, 70 per cent of them must also observe contract workers as 70 per cent of accidents happen with them.
10. Unsafe conditions are the result of unsafe behaviours, address unsafe behaviours on the spot.
11. People learn in four ways, observing, thinking, feeling, and doing. Observers must use all these four ways to make feedback effective.
12. In this organization, a lady officer from quality department is the only lady BBS observer in India.
13. Before the launch of BBS in the organization, the duties of BBS team/coordinators were discussed and concluded as below:
    - Distribute: Distribute checklist to observers daily in the morning.

- Collect: collect filled-in checklist from observers in the evening.
- Analyze: Analyze data weekly to understand the percentage of safe/unsafe behaviours.
- Display: Display a graph of safe/unsafe behaviours.
- Conduct: Conduct weekly meeting with all observers.
- Make: Make a weekly round in the plant together with all observers.
- Follow-up: Monthly review meeting of all observers.
- Report: BBS team would report monthly progress to HSE Head.
- Launch: Ceremonial launch of BBS.

4.8 CASE OF A HEAVY ENGINEERING MULTINATIONAL ORGANIZATION

During January 2010, a group of 22 safety cocoordinators of a heavy engineering organization employing more than 5000 employees in Pune underwent three days of BBS training. They reported that four unsafe behaviours, on an average, per employee, and decided that each safety coordinator would adopt one department and create BBS awareness among its people. They emphasized on the role and responsibility of each observer which is a backbone of BBS project. Newly trained observers were in the habit of observing unsafe behaviours and not giving feedback due to the old mindset. It was realized that the observer should not give feedback even in noisy work areas, take an observee to a non-noisy area so that the communication is heard fully. Though an observer observes and gives feedback to one worker at a time, he must not ignore many others who are not wearing PPE; his job is to point out to others as well when he is on his observation tour. An observer is not supposed to even ask the name of an observee during BOFP. Observers must assert and persuade safe behaviours on observees on the spot.

Observer's duty also is to close/follow-up unsafe conditions which he has observed during observation tour. If these unsafe conditions are beyond the purview of an observer, he has to contact engineering / maintenance department. Each observer may carry a digital camera to capture unsafe conditions for internal discussion during observers' meeting.

Benefits of BBS practical training reported by BBS observers:

1. Coaching and feedback skills of observers improved.
2. Deviations from SOP on part of observees corrected.
3. Knowledge of observees about unsafe conditions increased.
4. Knowledge of observers and observees about safety enhanced.
5. Mindset of observers changed.
6. Observation time reduced as observers maintained BOFP in routine.
7. Observees became more alert.
8. Observer's confidence and awareness increased.
9. Operators around observees got alert.
10. Relations between observers and observees improved.
11. Self-satisfaction of observers gone up. "We are doing something different.
12. Trust of observees in BOFP increased.

The observers discussed how at-risk/unsafe behaviours can be correlated with monetary costs of the injuries? This could be done by way of calculating the cost of injuries, man-days lost, property damage, court case, compensations, medical expenses incurred in the years previous to introducing BBS which could be compared with such costs saved in the years after introducing BBS.

4.9 CASE OF A MULTINATIONAL CHEMICAL UNIT: UNSAFE BEHAVIOURS

In February 2010, a group of 20 participants at the middle management level of a multinational chemical unit in Hyderabad participated in a BBS workshop and they observed that there are 71 unsafe behaviours of the operators working in the plant and on an average there are 3.5 unsafe behaviours observed per employee (71/20= 3.5) which meant that there are a total of 217 unsafe behaviours (62 employees × 3.5) existing at present. "*Since 2008, we were reluctant to start BBS, thinking that it's one more job, however we started half-heartedly, now going through some more literature and training, we got confident that we can do it,*" said a plant manager during this during the BBS training programme. "*BBS would give 100% safety without additional efforts,*" the company's pproduction head said.

4.10 CASE OF AN INDIAN CHEMICAL GROUP OF COMPANIES

In April 2010, 41 senior managers of an Indian chemical group of companies attended a day-long BBS workshop in Mumbai. In his opening remarks, an executive director of the group said, "*every person is a centre of safety.*" He added, "*if we wish to make a change in safety status, first let us educate our managers.*" The managers opined that some punishment or fear has to be there in improving safety scenario in organizations. An average of unsafe behaviours per employee observed by these managers during the last one year was found to be 5 (total unsafe behaviours 197/41=5) which created a requirement for BBS in their organizations. A manager realized that BBS needs to be linked with performance appraisal as observers contribute towards safety improvement of the organization through their initiatives and volunteerism of being an observer.

4.11 CASE STUDY OF A MULTINATIONAL FARM EQUIPMENT SECTOR

Eighty-one executives representing different departments in Maharashtra attended four one-day BBS workshops in July 2010. The Head of the plant mentioned that we had no accident in the past seven years, but lots of injuries in the plant happened. During OHSAS audit, it was highlighted that all officers and workers should be exposed to BBS. So the vice president took the decision to launch BBS in the organization.

The number of unsafe behaviours per employee observed by the executives during the preceding year was 9 which are multiplied by total number of employees, i.e., 2000 × 9 = 18,000.

The cost of per unsafe behaviour was calculated as per formula below:

Number of injuries last year × tentative cost per injury / Number of unsafe behaviours observed.

= 55 × Rs.10, 000 + (one major injury costing Rs. 2 laks) / 18,000 = Rs. 42 per unsafe behaviours.

It means when an organization has 18,000 unsafe behaviours in the plant, it surely needs BBS, and by introducing BBS in the organization, it could save not only employees from injuries but also the cost of Rs 7.5 laks per day caused by 18,000 unsafe behaviours.

The workshop participants discussed the following matters:

1. Four types of observees: slow learners, high-risk takers, rough-tough workers (these three types generally represent about 7 per cent of total number of workers), and the fourth ones are normally cooperative workers who are about 93 per cent. The 7 per cent require regular observation for reinforcing safe behaviours.
2. Creating awareness among employees alone does not ensure complete safety. Correction of unsafe behaviours is necessary. Though BBS focuses upon zero unsafe behaviour, it's equally important to create safety attitudes, beliefs, knowledge and practices among employees through regular safety education and training. BBS aims at creating safety culture beyond fixing zero accident/ incident targets.
3. Four steps of BBS implementation: one-day BBS awareness training to heads of all departments, two-day training to the trainers representing each department who would, in turn, impart awareness training to all employees; one-day training to BBS observers who volunteered from the awareness training groups of employees, one-day training to BBS steering committee volunteered from trained observers.

4.12 CASE STUDY OF A MULTINATIONAL AUTOMOBILE COMPANY

Twenty-three managers representing 10 departments of an automobile company in Maharashtra attended a one-day BBS workshop August 2010.

The number of unsafe behaviours per employee observed by the executives during the last one year was 6 which when multiplied by total number of employees, i.e., 1700 × 6 = 10,200.

The description of injuries occurred during the previous year in the organization:

4. 72 first-aid injuries × cost, i.e., Rs 10,000 per injury = Rs 7,20,000

2.34 non-reportable injuries × cost, i.e., Rs 5,000 per injury = Rs 1,70,0003.6. Reportable injuries × cost, i.e., Rs 10,000 per injury = Rs 60,000

Total number of injuries = 112; total cost of injuries: Rs. 9,50,000

Per injury cost = Rs. 8482

The cost of per unsafe behaviour was calculated as per formula below:

Number of injuries last year × tentative cost per injury / Number of unsafe behaviours observed.

=112 × Rs 8,482 / 10,200 = Rs 93 per unsafe behaviour.

The cost of unsafe behaviours is huge which differs from organization to organization depending on the number of injuries last year, tentative cost per injury, and the number of unsafe behaviours observed.

The BBS trained observer closes the unsafe behaviour there and then as it is observed on part of an observee during an observation tour.

4.13 A CASE OF AN OIL CORPORATION

"In hydrocarbon industry, one behavioural mistake can cost huge in terms of human loss and property loss, therefore safety has to come into our action," said a general manager.

Thirty-six heads of marketing departments across India attended two days workshop on behavioural safety during July-August 2010 in Mumbai. The unsafe/at-risk behaviours they observed in their workplaces are:

a. Safety shoes not being used by 30 per cent of employees.
b. Helmet not being used by 100 per cent of employees.
c. Gloves not being used by 100 per cent of employees.
d. Mobile at workplace being used by 70 per cent of employees.
e. Safety belt not used by 70 per cent of employees.
f. 30 per cent of employees did not attent safety committee meetings.
g. Smoking in prohibited area by 5 per cent of employees.
h. Safety goggles not being used by 100 per cent of employees.

They expressed that behavioural safety is a new concept and method for better safety management which is worth trying, it is implementable at grass-root levels, and is practical.

4.14 A CASE OF AN ENGINEERING PROCUREMENT CONSTRUCTION COMPANY

Eighteen senior management employees of an MNC company employing nearly 2000 employees participated in a day long behavioural safety workshop in August 2010. They observed on an average 5 unsafe behaviours in existing amongst their employees. So the total number of unsafe behaviours would be 10,000 for all employees.

On the office safety front, they found employees have six categories of unsafe behaviours involving: housekeeping (water spillage, unorganized files), use of equipment (computers / laptops), body positioning (bad posture), communication (talking loudly and disturbing others), visual focusing, and using mobile phones when at work.

They calculated cost of per unsafe behaviour as Rs 50/- using the following calculations:

50 first aid injuries @ Rs 1000/- per injury =	total Rs 50,000
3 lost time injuries @ Rs 20,000/- per injury =	total Rs 60,000
7 near-misses @ Rs 5000/- each =	total Rs 35,000
Total cost:	Rs 1,45,000

The participants observed that since these injuries and near misses happened near project completion time in last year, hence this may not provide the real picture. During the project, the total cost would normally be around Rs 5 laks. Hence the calculated cost of per unsafe behaviour is Rs 50/- (Rs 5 laks / 10,000 unsafe behaviours).

They prepared an action plan: steering committee (10 members), area BBS coordinators (20 members), and BBS observers (200 members).

4.15 CASE STUDY OF AN AUTOMOBILE ORGANIZATION: FARM SECTOR

Twenty middle and senior managers were trained on BBS in December 2010 in an automobile organization in Mumbai. They expressed that creating BBS teams in each work area is an easy way to implement BBS at workplaces. Each work area would customize BBS implementation as per their team's discussion. They used to observe and ignore unsafe behaviours, now they would observe systematically and provide feedback to the observee with a human touch. They would celebrate BBS day every three months. They learned that:

1. BBS takes us from safety awareness to individual alertness;
2. Their target changed from zero accidents to zero instances of unsafe behaviours.
3. It was a peer-to-peer safety movement and was a life saving process.
    - Human touch in conversation will work wonders; it changes the attitude towards safety.
    - BBS should extend to their vendors also.
    - People become concerned rather that casual or indifferent after training on BBS.
    - BBS team in each area would educate their workmen about the concepts of BBS.
    - They have 14,000 unsafe behaviours in the plant of 2000 total workforce which they plan to bring down using BOFP daily basis and also training more observers.
    - Listing unsafe behaviours for each job and displaying at workplaces would alert people.

4.16 BBS PROJECT LAUNCHED AT PATNA TERMINAL

The first Behaviour Based Safety (BBS) Project in Bihar State Office (BSO) was launched at the Patna Terminal recently. Blue-collared workmen, contractual workmen, TT drivers and security personnel were part of the project. BBS is a bottom-up approach where the responsibility of safety is taken up by the people handling day-to-day operations.

A team from head-office comprising Mr SK Singh, chief manager (HSE) and Mr Raj Kumar Dubey, chief T&D Manager, along with Mr SK Jha, chief terminal manager and his team of officers at Patna Terminal launched the project. Dr HL Kaila, an eminent psychologist and a BBS trainer from Mumbai, guided the entire project implementation.

In the first two days, the workmen were made aware about BBS and taken to various operating areas of the terminal like Tank Loading Filling shed, Pump House, Tank Truck parking, etc. They were asked to list out the safe and unsafe behaviours, which they observed during their round of the terminal.

During the programme, the workmen demonstrated how they had corrected 50 per cent of the unsafe behaviours on the spot. An action plan to correct other unsafe behaviours was also discussed by the workmen. Later, about 21 workmen, including the contract workmen volunteered to become BBS observers and take up the responsibility to daily observe safe and unsafe behaviours in the terminal. A steering team comprising six workmen and two officers was also formed to take this project forward so that all the unsafe behaviours could be corrected on a day-to-day basis. The enthusiastic workmen who attended the programme termed it "*Bhai bandhuon ki suraksha.*"

4.17 A FERTILIZER COMPANY IN MAHARASHTRA

Six groups of managers, workmen, union leaders were trained in behavioural safety approach for four days. The following observations were made:

Day	No. of participants	% of safe behaviour	% of at-risk behaviour	% of corrected at-risk behaviour	Employee type
1	21	48	52	41	Managers
2	22	56	44	42	Union leaders
3	20	60	40	40	Operators
4	18	73	27	57	Operators

1. The above table indicates that the percentage of safe behaviours has increased, percentage of at-risk behaviours has decreased and along with the percentage of corrected at-risk behaviour attempted by the trained BBS observers day by day.
2. The observers realized that the BBS project works at the individual informal level as well as it requires a formal structure in the organization. All HODs training in BBS and involvement is needed for its full implementation.
3. "*Production stream is disturbed as a result of any injury/ incident, hence correction of unsafe behaviours that triggers incidents shall boost production,*" a deputy general manager said.
4. They critically probed whether all 100 per cent accidents/ injuries are truly due to unsafe behaviours. Not really they discovered, as all plants have some unsafe conditions. Our behaviour has to be 100 per cent safe.
5. They revealed that all top-down safety systems/approaches are jammed now, and unable to reduce injuries/incidents, so it is time to go back to workers in order to create brothers of safety.
6. Every new project that is usually heavily dependent upon contract workers has experienced many fatalities due to at-risk behaviours caused by target achievement pressurized by supervisors.
7. Each trained observer revealed the confidence that they can daily save lives of fellow workers by correcting their at-risk behaviours.

4.18 A CASE STUDY ON BBS: JULY 2011

Twenty-five engineers participated in a day-long workshop on BBS. They belonged to twenty-five project sites for installing air conditioning and ventilation systems. The company in Chennai laid an emphasis on EHS.

They observed 192 at-risk behaviours in different categories as per the following table.

Behaviour categories	No. of At-risk behaviour	% of At risk behaviour
PPE: Using PPE, e.g., eye glasses, hearing protection, gloves, hard hat, etc.	62	32
Housekeeping: Work area maintained appropriately, e.g., trash and scrap picked up, no spills, walkways unobstructed, materials and tools organized…	20	12
Using tools and equipment: Using correct tools for the job, using tools properly, and tool in good condition.	38	20
Body positioning / protecting: Positioning / protecting body parts, e.g., avoiding line of fire, avoiding pinch points, etc.	12	05
Material handling: E.g., body mechanics while lifting, pushing and pulling, use of assist devices…	29	15
Communication: Verbal and non-verbal interactions that affect safety.	08	04
Following procedures: E.g., obtaining, complying with permits, following SOPs/SMPs /OCP/ SWP, lockout, tag-out procedures, etc.	08	04

Visual focusing (attentiveness):	05	03
Using mobile phones while working	10	05
Total:	192	100

Three areas specifically required regular observations by BBS observers about PPE, using tools and tackles, and material handling. They learned to react immediately to every unsafe behaviour as they observed it. They realized that BBS needs to be introduced to new contractors at the time of signing the contract. They observed that they shall implement BBS in spite of job pressure and deadlines, they have at their workplaces.

4.19 CASE STUDY OF AN OIL & GAS ORGANIZATION IN GUJARAT (MAY 2011)

A four-day exposure to 50 senior management employees on BBS approach and discussions brought out the following aspects. They represented refineries, pipelines and marketing terminals from all over Indian locations.

- 10 per cent of employees obey safety rules, 50 per cent sometimes follow, and 40 per cent don't obey or they lack safety awareness.
- Best part of BBS is that it is not instructive, and it comes from coworkers.
- How to deal with unmindfulness and overconfidence of operators. Observers observe the 'manifested behaviours of unmindful attitudes' which gets reflected into unsafe behaviour.
- Along with unsafe behaviour, observers may also observe the quality aspects of the product being produced by an observee.
- Senior managers felt that industrial relations (IR) issues may restrict implementation of BBS, so it is essential first to create awareness on the concepts of BBS to all employees to achieve a common mindset before implementation.
- They observed that, on an average, 6 unsafe behaviour per employee in their organization.

- Some of them felt that personal touch with employees really has worked wonders. "*The production increased from 800 to 1200 drums per month when I have been talking to my workmen about the well being of their children,*" said a manager.
- They named the BBS project as Employee Suraksha Scheme (ESS) and developed a monthly employee safety index (ESI) which donates to a total of unsafe behaviours minus a total of safe behaviours divided by a total of safe behaviours.
- They felt that they should be awarded with the MBBS (Masters of BBS) certificate after three days of BBS training. We felt confident to implement.
- "*To live safely, take an oath; follow an approach of BBS path,*" said a manager.
- For the success of BBS, three factors are critical: management engagement with BBS activities over a period of time, observer's ability to convince, and observee's acceptance of BBS. Awareness drives for all across organization helps achieve these objectives.
- It is very important to address all the questions of managers and workmen about the implementation of BBS within the organizational and cultural constraints.
- The insufficient leadership of top management towards organizational safety can pose impediments to BBS implementation.
- There are stages of change management in BBS project development. The employees would have apprehensions in the beginning, then gradually they would conform to the positive changes seen through BBS approach, and then they engage themselves to full BBS activity in the organization.
- Terms, such as, BBS Committee should be replaced with BBS Team, and BBS Audit with BBS Review as they have negative connotations.
- Internal/external reviews should include interaction with observers and steering team, understanding BBS monthly progress, and meeting the observees.

- Observers should not only care for employees' safety, they also look for plant safety.
- Checklist development is a major activity which clarifies the unsafe behaviours across work areas to be focused upon.

4.20 CASE OF AN OIL & GAS ORGANIZATION IN BIHAR (JUNE 2011)

Fifty-three officers and 100 operators, drivers, contract workers underwent a 7-day BBS training for its implementation in their work areas which brought out the following experiences:

1. They pointed out that in a location where they had the best safety systems and the safety index was 100, a major fire took place, and the location which was not considered to be much safe did not experience any such disaster or accidents.
2. Does BBS care for financial safety or misappropriation of funds? Does observer observe and correct such behaviours among their employees?
3. BBS is a support mechanism for existing safety systems.
4. Trained observers amongst contract labourers observed and corrected many unsafe behaviours such as two workers were sleeping under trucks, one driver was smoking in the parking area, and one driver speeded his truck in terminal area.
5. Operators become alert even by looking at the checklist and pen in the hands of observers.
6. Contract workers reported that they got knowledge, direction and experience of BBS approach.
7. Top-down safety approach depicted passive leadership for safety and the bottom-up approach from the active observers revealed an active involvement in safety.
8. Real experience of BBS lies in out-of-classroom training for its implementation in the field/work area. Both formal and informal (out-of-classroom) BBS training for management and non-management employees is necessary. Their queries and issues related to organizational culture

need to be addressed with openness as some managers may feel unconcerned about BBS applications due to organizational politics, or management's over-emphasis on production, or thinking that BBS would be an additional workload, etc. Informal meetings with management employees during tea/lunch breaks help clarify some such issues and help in BBS implementation.

9. The contractual labour BBS observers who can't read and write need to be allowed to '*observe and correct unsafe behaviours*.' They need not be necessarily involved in checklist filling. Such BBS observers have proved to be very useful in the process of observation and correction of unsafe behaviours. BBS approach makes use of each and every person in promotion of safe behaviours in an organization.
10. "*In the past, we trained workers but did not utilize them", we do safety as our job, when they do it, they feel elevated,*" said a safety manager.
11. "*What safety officer could not achieve in terms of safety compliance at workplace, we did it as we regularly remain with our workers,*" said a contractual worker.
12. "*We used to say, safety by all, BBS has made it happen in true sense,*" said a manager.
13. BBS observers are like radars, always alert.
14. The observers estimated behavioural safety alertness of workers in the organization as 66 per cent, which meant that every third behaviour of workers is unsafe.
15. BBS provides all employees a status of equality in terms of safety.
16. Most management employees do not put a hand on the dusty shoulder of workers.
17. BBS success lies in no reporting to management or no interference from management, the BBS steering team monitors the BBS project in an organization.
18. Broadly, BBS observers care for health, safety, and environmental aspects.

4.21 CASE OF AN AIR FUELING STATION (AUGUST 2011)

A three-day BBS training and workshop was conducted at AFS Kolkata during 18-20 August, 2011.

A heterogeneous mix of eighty persons that included officers, staffs memebers, contractual employees, and security personnel, participated in the training for implementation of BBS at the location.

The incumbents thus being trained in BBS, not only acquired the necessary skill of marking unsafe behaviours, but also become instrumental in correcting the observed unsafe behaviour by the way of demonstrating the correct form of safe behaviours which is evident in the table below.

Group	Safe behaviour	At risk behaviour	Corrected behaviour
1	54%	46%	66%
2	67%	33%	52%
3	63%	37%	45
4	68%	32%	54%

A leap of 14% is observed to occur in two days as far as the tendency of "safe behaviour" is concerned which was mere 54% in the beginning and leaped at healthy 68% in two days.

During implementation of BBS, it was realized that the resistance has been converted to reception among the employees.

4.22 A PUBLIC PROGRAMME ON BBS

Twenty-six personnel from management cadre from 12 organizations participated in a one-day open programme in Gujarat. They realized that: BBS is a 'near miss to never miss an unsafe behaviour' project. It's a cultural change and BBS in safety is like Bournvita in milk. BBS is a spiritual way of dealing with safety in an organization. It's a "*surksha mitr abhiyan.*" It is a *shouting to soft* approach. The circle of safety is incomplete without contractual workers, and in BBS, we include everyone who enters the campus as it is always a company-wide project.

4.23 BBS IMPLEMENTATION EXPERIENCE OF MANAGERS

Based on the feedback and our visit to the locations where we had implemented BBS, we found it had to be effective. For any new programme, there are apprehensions, now the results have been very encouraging, plant safety has improved. Earlier we had withdrawn from the workmen, we had ignored them, and we had not taken from them. We have to go to their standard and their body language. If we are asking them to work for more number of hours, it would essentially mean that we need to care for them more for their safety. Making a shift from traditional safety approach to BBS paradigm gives a sense of joy. We need to care for workmen consistently and keep the process of dialogue on. Recognition of observers in front of others is important to motivate them. Public reprimand must be replaced by public praise. In overall, the BBS approach provides a feeling of empowerment, ownership, informality, discipline, pride, empathy, and synchronizes the potential for the benefit of the organization. Pre-launch understanding of BBS is very significant in its success in the organization. Within a period of five months, our observers identified and rectified 75% of unsafe conditions at the worksite.

4.24 SAFE BEHAVIOURS INCREASED IN THREE DAYS

Three batches of fifty eight operators were trained as BBS observers in a large engineering organization in Pune in December 2011. Having gone through the BBS training, they observed their colleagues at three shops, of which the results as under:

Day / batch	% of safe behaviours	% of at-risk behaviours	% of at-risk behaviours corrected	% of at-risk behaviours to be corrected
1	59	41	45	55
2	68	32	55	45
3	74	26	42	58

The percentage of safe behaviours increased from 59 to 74%, at-risk behaviours decreased from 41 to 26%, and they also corrected the at-risk behaviours from 42 to 55%. In addition they observed that most of the eye injuries took place to the visitors/ supervisors at the shop floor due to the sparks of welding operations as they did not wear safety goggles. An operator asked an observer during his observation tour, "*has the boss asked you to observe me*," to which the observer replied, "*no, it's because I am concerned about you*," the operator then said, "*then it's ok*."

4.25 SAVING OTHERS IS A HUMAN INSTINCT

In a BBS workshop with 26 top management executives of a large construction company in Mumbai, a vice president remarked that saving others is instinctive, people would normally extend a helping hand when somebody is about to fall. Hence, brother to brother safety (BBS) generally comes naturally to all of us as per behavioural safety perspective. And unsafe behaviours happen so naturally and intentionally to almost all of us, thus we all need big brothers of safety (BBS) around us.

"*An aggressive emphasis by two of our new board members has given a re-focus on the safety movement as there were many fatalities across 30 sites across India*", an EHS head said. "*We shall keep BBS on our board meeting agenda, its status will be reviewed, we shall pen it down and circulate to all employees*", the managing director asserted.

4.26 BBS IS A DYNAMIC IMPLEMENTATION PROCESS – A CASE (JANUARY 2012)

In the state of Rajasthan, two days of BBS training was provided at a cement plant to 92 senior and middle managers representing ten departments, such as, civil, mechanical, instrumentation, packing, power, stores, mines, QC, electrical, and production. The senior vice president (production) remarked that efficiency parameter has now included safety dimension also. One is efficient if he is working safe in the plant. The participants observed the following during deliberations:

1. Unsafe behviour and safe conditions are generally positively correlated as people tend to behave unsafe and less alert as they perceive conditions of work around are safe.

2. Safety and production are positively correlated. When people behave safe at work, they do not experience injury or accidents which in turn saves injury-related production loss or slow down.
3. BBS is more a dynamic implementation process for observers than making it as system of compliance for maintaining documentation.

4.27 DIFFERENTIAL PERCEPTION OF UNSAFE BEHAVIOURS IN A FERTILIZER COMPANY

It is found from the table below that the middle level managers/ union leaders observed less number of safe behaviours as compared to the senior managers. This means that the risk perception varies among people in an organization.

	% of safe behaviours	% of at-risk behaviours	% of at-risk behaviours corrected	% of at-risk behaviours to be corrected
35 Senior managers	70	30	73	27
44 Middle level managers/union leaders	58	42	56	44

4.28 FIELD OBSERVATIONS DURING BBS ACTION RESEARCH

1. BBS implementation has reduced average unsafe behaviour of employees from six to two from 2009 to 2011 at Bayer's Crop Science in Gujarat. But SOP has been changed and not communicated to us, we are simply asked to carry it out which reflects that officers have not changed much even after two years. The use of PPE in our company has improved so much so that people started bringing helmet to the canteen and we had to arrange a table to keep helmets.
2. Each of the BBS training group of 25-30 people has demonstrated almost 50% of safe and 50% of unsafe

behaviours as being experienced regularly at home, on the way and at the plant. Hence, it is felt that almost everybody carries a set of unsafe behaviours which are an integral part of their lives. Thus, these individual unsafe behaviours are a kind of RDX for accidents and a source of terrorism against safety.

3. It is perceived that conversion of unsafe behaviours into safe behaviours requires four to five interactions between an observer and observee.
4. BBS checklist is not filled-in so mechanically that the observer neglects the main purpose of BBS which is primarily to identify and control at-risk behaviour so that the observee is saved from risk. A BBS observer must save any observee following unsafe practices and should not go on filling checklist mechanically. In this case, first save then document.
5. The education of BBS to employees can be useful at home, on the way, and at the plant.
6. An observer need not find out "reason" on why a person is engaged in an unsafe behaviour or why an observee "should" not behave safe, rather an observer has to actively observe and give feedback showing his concern to an observee.
7. "Observation schedule" is very important for each observer so that he can cover all locations/ shifts/ observees/ each working hour during his daily observation tour.
8. During BBS meetings, never use the name of an observee in presence of others.
9. There are two types of observers, one who gives concession, that is, he observes unsafe behaviour of an observee but does not document; the other is the one who meticulously documents all at-risk behaviours and gives feedback. The latter is a better option.
10. All observer groups must be given some incentives, e.g., individual recognition, group accolade, etc., as they participate in bringing down at-risk behaviours and creating safe environment.

11. A plant head said very enthusiastically, "*Through BBS training, we wish to turn everybody into safety officers. People need to take care of their own safety as well as others.*"
12. How feedback between observer and observee works in BOFP: it follows stages such as informing observee about unsafe and safe behaviours, observee denies and feels embarrassed about his unsafe behaviours, then he experiences alertness, consciousness, slowly he attempts to change his unsafe behaviours to safe behaviours, then realizes internalization of such behaviours, stabilizing change, self-observation, and is then ready to change others.
13. The workers should not feel that management is dumping safety on them by launching BBS as it is a bottom-up approach, they should not get a feeling that management is shirking away from their safety responsibilities towards workers.
14. Suppose a company implemented BBS and after six months of its implementation, an accident occurs. The BBS observers may suddenly feel that they have failed in their duty. The management may feel that we have given control of safety in the hands of workers, it is their problem. At this time, we should first look at the percentage of unsafe behaviours across different units because it is the root cause of any near-miss, injury or accident.

4.29 CONCLUSION

The present survey on organizational case study researches reveals that the unsafe behaviours do exist across Indian organizations and employees engage in unsafe behaviours on daily basis.

Managements in India have started believing that unsafe behaviours need to be controlled in order to ensure total safety at workplaces. Engineering controls alone do not provide adequate safe workplace unless behavioural safety is practised (Krause, 1995).

Though OHSAS 18001:2007 has included three clauses as mentioned above that emphasize behavioural aspects of safety, the organizations have yet not followed it exactly as the OHSAS 18001:2007 does not provide any guidelines on how to implement these clauses. However, Indian organizations have started considering human behaviour aspects of safety at workplaces more as compared to yesteryears.

It is also observed that the BBS training facilitates increase in number of safe behaviours, helps reduce number of unsafe behaviours and also assists in reducing the number of unsafe conditions in the organization.

BBS is all about involving people across departments in an organization as a bottom-up approach (Locke, et al., 1998). In order to implement BBS at workplace and create safety culture, the following steps are recommended:

1. Organize an awareness programme on BBS for management staff at all levels.
2. Conduct awareness training of employees across the plant;
3. Select (20%) observers from BBS trained employees; and form steering committee from BBS trained employees.
4. Carry out in-plant practical training of BBS observers and steering committee members on how to set up the observation process, how to develop the measure, making accuracy and consistency checks, steering committee functioning, etc. At this stage BBS observers and steering committee members are fully prepared to implement BBS at workplace.

4.30 SUMMARY OF BEHAVIOUR BASED SAFETY

1. Root cause behind an accident/fatality is unsafe behaviour.
2. Unsafe behaviour can be practised by anybody.
3. Any unsafe behaviour cannot be ignored.
4. Alert the person as and when he performs unsafe behaviour.
5. Do not ignore any unsafe behaviour until it is corrected.
6. Unsafe behaviour comes both naturally and unintentionally.
7. Do not stop work, stop unsafe behaviour.

8. Production is not hampered by following safety regulations.
9. Unsafe behaviour and safe condition have a positive correlation. People tend to behave unsafe when they see environment is safe, e.g., they increase the speed when they perceive that highway is free and totally safe.
10. BBS is no common science. It is based on applied behaviour science.
11. World over, 23 laks of people die every year because of workplace accidents.
12. 70-80% persons targeted by unsafe acts are contractual workers.
13. BBS is an "OHSAS 18001" compliance.
14. Behaviour is definable, observable, correctable, measurable, and implementable.
15. BBS is the art of safe living.

5

101 Varied Reflections on Behavioural Safety

This part of the book shares a qualitative / narrative data of the national action research survey on behaviour based safety in India. It is an ongoing longitudinal study in which the BBS training was provided to nearly 9300 workmen (operators, technicians, contractors) and union representatives and about 1800 managers at all levels between 1997 and 2011 in diverse multinational organizations across industrial sectors (such as, oil and gas, power, heavy engineering, automobile, chemical, coal, construction, pharmaceuticals, electrical, nuclear, etc). They were exposed to the BBS concepts; observation and feedback processes; and, implementation of BBS in an organization. The training focused on providing an indepth sense of understanding and application of the concept and process of behaviour based safety concerned about correcting unsafe behaviours for reduction of accidents and promoting safe behaviours for developing injury-free cultures in their organizations. The workshops were held in both Hindi and English languages with 30-50 members per batch.

Struggle to manage safety situation at workplaces continues. Audits are done, awards have been received, and documents are in place. Still organizations wonder what to do to ensure safety of people. "An injury hurts the organization because a factory inspector comes and investigates and asks so many questions," a general manager stated.

"*Section 111 of the Factories Act has already emphasized legal requirement for behaviour much before OHSAS*," said a deputy chief inspector of factories in Andhra Pradesh. He added that an unsafe worker can create accident in a safe environment and also a safe worker can prevent accident in an unsafe environment. As an individual, you are responsible to discharge your duty to see safety of your workers.

BBS is also therefore referred as *Become Brothers of Safety* to save lives of people at the workplace.

The reflections and narratives of workmen about BBS across India produced the following 26 themes which would hopefully enrich the existing knowledge of behavioural safety:

1. Accidents/injury don't spare anyone, even managers.
2. Four significant aspects of BBS.
3. Annotations on safe and unsafe/at-risk behaviours.
4. Conceptual extensions of BBS.
5. Shared perspectives on BBS approach.
6. Vital characteristics of BBS observers.
7. Positive changes attributed to BBS.
8. Limitations/deviations on organizational safety culture.
9. BBS and other safety systems.
10. Managerial perspectives on BBS.
11. BBS and minimum standards of safety.
12. OHSAS 18001 and BBS.
13. Relevant questions on BBS.
14. BBS is beneficial.
15. BBS is a request for safe behaviour.
16. Awards can make people complacent.
17. Voices of Indian organizations.
18. Unsafe behaviours identified by BBS trained observers.
19. BBS project deliberations.
20. How to scale-up BBS projects.
21. Family perspective on BBS.
22. Safety delays production.
23. Ignoring unsafe behaviour means approving it.
24. Each unsafe behaviour is a risk for the entire plant.
25. Total safety culture.
26. Conclusions.

5.1 ACCIDENTS/INJURY DON'T SPARE ANYONE, EVEN MANAGERS

1. Unsafe behaviour can happen regardless of your position, education, experience and age. A vice-president went up on the fourth floor to inspect a construction project, he received a call on his mobile and started talking, got so

engrossed that he just put his step forward and fell to the ground and died on the spot.

2. An engineer on the shop floor thought of crossing a conveyor belt while it was stopped, as he crossed, it started moving, he was crushed to death.
3. A deputy general manager got a serious eye injury when he was observing a workman without wearing safety goggles and an object flew from the machine and hit him in the eye.

"*Actually, each unsafe behaviour has caused a fatal accident at some time or another*", a deputy chief engineer said. When we are working, our attention should be on safe working.

5.2 FOUR SIGNIFICANT ASPECTS OF BBS

There are four aspects of BBS: observation of unsafe behaviour, correction of the same, there and then, in your presence. Examples: "*I noticed, a person was clicking photographs while climbing an escalator I thought of telling him that this is an unsafe behaviour, in the meantime, he turned backward and took another photograph, by the time, the stairs got over, and he fell and got a head injury. If I had corrected him there and then, he might have saved himself*" – head, ONGC Uran plant.

Similarly, "*three young people went on the height to work, all three were wearing safety belts, but two had hooked it but not the third person. While working, all three fell, two got saved but the third person fell and died as he had not hooked his belt. In case the other two had observed and informed him to hook the belt, it might have saved his life,*" – Head of L&T plant at Hazira.

5.3 ANNOTATIONS ON SAFE AND UNSAFE/AT-RISK BEHAVIOURS

Workmen engage in unsafe behaviours because of three reasons: PPE not comfortable, PPE not provided and production pressures by supervisors.

4. On an average 4 at-risk behaviours per employee are observed in the organizations. Three or four unsafe behaviours when combined trigger accident or injury.

5. Unsafe behaviours are related to an individual and unsafe conditions are related to a work environment.
6. Unmanned organizational unsafe behaviour: though there are many observers around, there is no one to check a process. For example, when filled LPG cylinders were being passed through water to check gas leakage, the process was unattended.
7. In high-risk areas, operators observe safe behaviours more than the less-risk areas.
8. The unsafe behaviours observed relating to: using mobile (30%), non-use of PPE (20%), not using right tools (10%), bad housekeeping (30%), complaining of pending safety conditions (20%), employees taking safety casually (30%), not keeping hygienic conditions (30%), feeling pride in violating safety norms (30%).
9. Is BBS only a correction of unsafe behaviours, is it preventive approach? It can become preventive only if we make BBS a data-based and the data of unsafe/safe behaviours in organizations would decide whether we are preventing the same.
10. Unsafe behaviours observed by workmen and managers are different. Workmen observe more minutely as they belong to the shop floor.
11. Don't delay in giving feedback on unsafe behaviour, otherwise an accident or injury might take place.
12. There are three ways to change behaviour from unsafe to safe: 40-50% by enforcement, 10-15% by safety education and training, and about 50% by BBS approach.

Being a safety conscious company and having safe behaviours among employees are two separate phenomena. Sometimes workmen succumb to an unsafe behaviour to complete the work under pressure. People behave safe or unsafe as per their risk perception not actual risk present there. Its management's unsafe behaviour when they don't provide proper equipment to workmen or modify unsafe conditions for a long time. Safe behaviour should become core value of business in organizations. "*an unsafe behaviour of a single person leads to loss of myself, my friends and my plant*" – a shopfloor operator.

5.4 CONCEPTUAL EXTENSIONS OF BBS

13. "*In others' safety, lies my own, I understood this through BBS approach*" – a workman.
14. "*BBS is the basic to broad safety*", a union leader stated. Union people are never against the safety procedure.
15. "*Housekeeping means where everything is in place*" a workman.
16. BBS is the backbone of workplace safety.
17. Though safety is an organizational issue; BBS makes it an aspect of self and a life issue.
18. BBS is doing something for someone.
19. A lady observer said, "*BBS means together we can do it.*"
20. 'Behaviour Observation and Feedback Process' is regularly observing and communicating the unsafe behaviours to employees. What is communicated regularly gets achieved.
21. Include BBS vaccine into induction training of new trainees and contract workmen.
22. "*BBS is like homeopathic medicine, it's sweet and slow. Each worker becomes an ambassador of safety*", director of a safety institute.
23. BBS is bottom to board safety. It's also breaking the barriers of safety.
24. BBS is breaking the barriers of safety. It should also be introduced to schoolchildren for peer-to-peer safety.

Industrial safety is all about auditing, reporting, investigation, training, process safety, etc., behavioural safety is all about observing and correcting unsafe behaviour and promoting safe behaviours. BBS actually removes the '*un*' from un-safe. It implies 24-hour safety. "*If you observe anybody behaving unsafe, you need to save him immediately, beyond which you may not get time to save him. I failed to alert one person as I was about to tell him, before that he turned and fell*", said a plant head.

5.5 SHARED PERSPECTIVES ON BBS APPROACH

25. An organization named their BBS project as S-Care (safe care, not scare).

26. BBS is preventive maintenance. It is implementation of safe operations (ISO).
27. In some organizations, BBS is manager-driven and is supervision-based. In principle, it's a bottom-up approach.
28. Supervisors/site incharge normally pressurize workmen to work without much safe practices. All immediate supervisors must be trained in BBS.
29. Toolbox talk does not cover all workmen, as only interested ones attend to it, whereas BBS emphasizes one-on-one observation on regular basis at all sites.
30. No work areas are actually safe as far as project sites are concerned. In these situations, safe behaviour is the only option for workmen safety; hence it's necessary to alert each other for safe behaving.
31. Online BBS programme can be useful for creating awareness among employees through intranet within organizations.
32. BBS gives a 'medium to connect' with people whom you did not talk earlier in the plant.
33. BBS is democratic, positive and human approach to safety. It is 'safety with interpersonal skill'. It's mutually educative between observer and observee.
34. Safe culture is 'safe behaving' of people in the organization which can be measured through BOFP by observers on monthly basis.
35. BBS is a priceless tool kit. It means best behaviour supply (BBS).
36. "The satisfaction of saving lives is greater than getting crores of rupees," – a fire engineer.

Safety efforts should boost production; not drag it. Mahatma Gandhi did not resort to violence but achieved his objective; BBS observer does not conflict with production but achieves safety. Aggressiveness and harsh argument with observees are contrary to BBS approach. According to an observer, "*Earlier we abused and dealt with workmen harshly, after attending BBS workshop, we would talk to them with patience and without arrogance*." Jesus Christ sent twelve of his disciples/saints across the world to spread

Christianity; a small team of active observers can spread BBS awareness across organizations.

5.6 VITAL CHARACTERISTICS OF BBS OBSERVERS

37. "*A BBS observer is a teacher for workmen and he teaches in their language*" – a safety marshal.
38. A safety marshal must set aside his ego while interacting with employees.
39. Observers must watch out whether any work activity is left without observation.
40. Observer would take an observee to the occupational health centre if he is having any health problem, this way he should attend to safety and health aspects of observee.
41. Observer must teach observee how to preserve PPE, not to rough-use them.
42. Before starting on an observation, observer must broadly look around the work area and then decide what to observe, is there any unsafe behaviour going on?
43. It is better to refer to observers as *behavioural safety counsellors* (BSCs) as observation is not considered in positive sense by the people in organizations.
44. Sometimes not only observees are rigid for behavioural change, the observers also may be rigid. Observer's rigidity can be overcome as he matures with BOFP.
45. It is preferable to have a BBS core committee for leading the BBS project and also a BBS steering committee of observers for daily practical coordination.
46. "*BBS observers performance needs to be linked to their individual performance appraisal and also to reward system*", a manager.
47. Each observer generates 'a sense of safety' in each observee.
48. Observer has to be disciplined and basically strong as making wooden furniture like sofa-set, its frame has to be strong.
49. Workmen think that PPE are are like extra burden on the body, but workmen have wear them despite discomfort for protecting their body, normally this is the clarification

given by observer to the observee who finds fault in using PPE.

50. Observers see that BBS approach does not conflict with production targets.
51. "*After observations I realized that an observer was hidden inside me and I found certain unsafe behaviours of an observee which I was also doing*", an observer said.
52. BBS observers encourage observees to comply with safety rules and regulations.
53. Basically, observers not only alert observee, they also enquire into the passive safety processes or systems and alert them.
54. Select observers who are passionate and aware about safety in the organization.
55. Observers close unsafe behaviours there and then in their presence otherwise they might find workmen injured on return if left with unsafe behaviours. So observers take now-do attitude rather will-do later.
56. Observers allocate responsibility of work areas for BOFP and also for creating awareness among workmen within the stipulated time frame.
57. In an organization, all observers took safety pledge and signed it and displayed it in the plant. They stated that our mindset is enlightened, path is clear for excellence in the plant. But change is not a one-day process, it takes long time, we have to struggle, we have to face with patience.
58. An observer is like an LIC agent who never gives up but achieves results.
59. Observers can also check the quality of products along BOFP.
60. Observers comments: "*I am proud not because of my observation but due to the fact that I closed unsafe behaviour*". "*We can sustain the behavioural change if we regularize observations*."

Pandit Jawaharlal Nehru, the first Prime Minister of India said, "*My air force is always ready like the fire brigade*". BBS observers are also like fire brigade, 24 × 7 instant observers and corrector of unsafe behaviours.

5.7 POSITIVE CHANGES ATTRIBUTED TO BBS

61. "*BBS training has more or less resolved my apprehension that implementing safety standards in practice is difficult. The training has shown that politely and consistently asserting safety in practice would bring positive change*", a workman.
62. "*There is a drastic change after BBS launch in terms of increased time in safety discussions, wearing PPE, up-trend of safe behaviours*", BBS observers.
63. Following are the differences noticed post-launch of BBS by twenty-one observers: mostly workmen wore helmets, housekeeping maintained, change in oneself, safety guards used by workmen, earthing was done, tank-truck drivers alerted others to wear PPE, workmen wore goggles, contractors promised to arrange PPE for their workmen.
64. A safety officer said, "*I am happy that today I have got 12 trained observers with me, I feel stronger. I was feeling alone as safety officer in the organization.*"
65. "*Earlier we were five people making rounds, now we are 40 people after BBS training*", a safety head.
66. "*TPM educated us about near-misses, but BBS taught us about unsafe behaviour,*" a safety coordinator.

After having gone through BBS workshop, the observers found the following changes within themselves: increased confidence to interact, a better way to talk with workmen, sweetness in interaction, an emotional touch, increased observation power (macro to micro) and scope (limited to unlimited). They also found changes in observees/employees, such as, increased safety alertness and awareness, rectification of unsafe behaviours, wrong practices reduced. The trained observers planned to introduce BBS in daily tool box talk.

Limitations / Deviations on Organizational Safety culture

67. "*Top managements have initiated BBS for getting orders from foreign clients. Accidents still keep happening after*

BBS implementation. The workmen who implement BBS, they forget within 2/3 days or immediate supervisors focus on production and bypass safety", a safety engineer.

68. In a construction project activity, 25 persons were identified (by a safety officer) who were working on a height without work-permit.
69. "*Safety emphasis gets dissolved as it goes down from the main contractor to next level of sub-contractors*", a general manager.
70. "*Indian organizations could not develop safety culture, still we are enforcing, if we don't go to the site for 2/3 days, we still find lapses, we find people without PPE*", a safety professional.
71. In projects, people are more concerned about completing work by passing safety.
72. "*Contract workmen need to be told once or twice but our regular employees need to be told six time to follow safety practices at work,*" a safety manager.
73. "*When it's an audit time, then only I wear helmet*", a workmen.
74. More resistance actually comes from regular employees than contract workers in implementing BBS. "*Safety officer is coming now, I am going to tell ten things which are pending*", this is the reaction of regular employees, said a safety officer.
75. "*As safety professional, we shout and irritate, somewhere we are doing mistake, we should tackle with patience and positive approach*", Head of HSE, DrReddys Lab.
76. Work permit is considered a permit to work, and not to read.
77. Culture of fear and distrust at the shop floor resist change among employees, it's a barrier in BBS.

In India, human resource is considered a commodity by the employer, one will go, and another will come. Unless we value human life, safety does not become a core value of business. "*Safety is still considered with money considerations in organizations and its status is like a daughter-in-law in the typical Indian family,*" a safety manager.

5.8 BBS AND OTHER SAFETY SYSTEMS

78. BBS and STOP are similar in principle but the structure is different in terms of observers' profile, checklist customization, and the psychological emphasis.
79. Safety records are not sufficient to make plant safe.
80. BBS is alertness-based; other safety systems are awareness-based and top-driven.

Some workmen hold on to negative approach and talk negative about existing safety systems in the organization. They gradually open up when constantly explained with positive approach.

5.9 MANAGERIAL PERSPECTIVES ON BBS

81. Managers need to motivate observers daily for BOFP. Ten minutes of BOFP makes all the difference in safety of an organization.
82. A senior manager said, "*Each workman should not only go home safe but also clean*".
83. BBS needs to be included into toolbox talk, induction training as well as regular safety education and training programmes.
84. Top management is not exempted from BBS; actually top manager is the chief observer.
85. "*Along with BBS, we also need to buy better quality tools*", vice-president (operations).
86. BBS needs to begin from top person of the organization, and then only it would be successful, it requires strong leadership back-up.
87. Maximum behavioural change can be achieved by enforcement, training and BBS.
88. Caring is more important than enforcement and education.
89. Improve communication with politeness to get acceptance of safety implementation. No job is so important that it can be done unsafely.
90. Training to observers and management is the key to the success of BBS.

91. Lack of management support can lead to frustration among observers when proper quality of PPE or tools is not provided despite many reminders.
92. "*Let's follow BBS at the shopfloor in the name of humanity*", a manager.
93. "*The concept of public sector has failed; we don't want to fail in BBS. Target of observers did not meet, it does not mean that democracy does not work and only dictatorship works,*" vice president (operations).

Sometimes psychological resistance for BBS comes from managers. A manager said, "*BBS means a big-big stone which refers to a barrier created by managers in introducing it. Safety leadership poses complexity when safety issues are not addressed by plant manager.*" 'Misalign with management' is the fear, so the line managers bypass safety, said a senior HSE professional. Formal BBS policy and top management's commitment are required for BBS success. Safety leadership must take off from the workmen level as managers come and go. Advertisement for BBS and reward for its observers are very important for its success.

5.10 BBS AND MINIMUM STANDARDS OF SAFETY

BBS is more required and useful when minimum standards of safety are not available in an organization. Observers alert workmen about their unsafe behaviours. It does not mean that minimum standards of safety are not required when BBS is introduced.

5.11 OHSAS 18001 AND BBS

OHSAS is all about occupational health and safety, BBS is the only clause that covers behaviour science when 90% accidents are due to unsafe human acts. Some organizations introduce BBS only for OHSAS compliance. "*OHSAS is merely documentation,*" a safety manager.

5.12 RELEVANT QUESTIONS ON BBS

94. We launched a safety persuasion approach in our township for people to wear helmet while driving, it did not work, how BBS would work? BBS is a 'systematic training and data-driven approach based on BOFP' which was perhaps not used in your approach.
95. Contract workmen are floating workforce, how to cover them under BBS? We also need to essentially train observers from amongst the regular contract workmen. The regular employees must also observe contract workmen.
96. Even with BBS approach, most of the workmen go unobserved most of the times at work. Then what? Essentially and obviously, we need more observers, but observee also spread the message of BBS to other workmen.
97. "*What's the key to plant safety is the behavioural safety*?", asked vice president of a heavy engineering plant. He further added that the earlier QC department was considered to be responsible for quality control but then QC was made as a line function. Similarly, BBS is now a bottom-up approach.
98. How critical is the unsafe behaviour? Each of the unsafe behaviours has already resulted in fatal. So any unsafe behaviour can be critical.
99. Observers ask, "should we mark behaviour as safe after having it corrected?" Yes, as a motivation to the observee. No, unless followed and seen as maintained as safe.
100. Can observers address personal/family issues of an observee? Yes if an observer feels experienced enough to deal with or he should refer the case to a professional counsellor if an observee is alcoholic, drug addict, or having marital discord. Observers can also be further trained in counselling skills.

5.13 BBS IS BENEFICIAL

The workmen commented that being safe in the plant means being alert all the time. Any external alertness makes a difference in behaviour. If we are behaviourally perfect, safety is attached to it. Educate workmen in their regional languages. Injuries in our plant have come down to 32 from 102 within six months of implementing BBS. Due to BBS intervention, the number of safe behaviours is increasing everyday/shift. Number of unsafe behaviours is decreasing everyday/shift. Number of unsafe conditions is decreasing everyday /shift. BBS provides scope for interactive communication with sensitivity for workmen.

5.14 BBS IS A REQUEST FOR SAFE BEHAVIOUR

"*When I was driving inside the plant, an observer (a worker) came to stop my car and requested to slow down the speed saying that this month is for speed control of vehicles within the plant. I felt so nice about this way of changing to safe behaviour through BBS approach*," a senior manager.

5.15 AWARDS CAN MAKE PEOPLE COMPLACENT

According to a general manager, "*Sometimes getting more number of awards would make employees lethargic about safety*."

5.16 VOICES OF INDIAN ORGANIZATIONS (855 PARTICIPANTS FROM 21 ORGANIZATIONS)

- I have become instant social worker with BBS training. It's redefining safe work environment. BBS is an 'art of safe living'. I will save my colleagues as his unsafe behaviours can affect me also. Other is not other, he is our divine brother.
- Do not stop work, stop unsafe behaviour to reduce accident or injury. Accidents are triggered by sequence of unsafe behaviours. "*A small mistake of any person can lead to damage to society, most accidents happen when we are*

sluggish in the morning hours or when we are in a hurry to go in the evening hours," plant head of RIL Kakinada.

- Why do we excuse ourselves from correcting unsafe behaviour? BBS is the kaizen of safety. Cost-saving from accidents or injuries must be the prime focus of BBS.
- Indian management takes a view that BBS is a low cost safety management device; hence they can do away with safety training or providing PPE to employees. Mock-drill has become '*chai-nashta*' in organizations, people don't actively participate.
- "Motivation gift must be given to active observers every month," said an operator. Observers have a ready and rapid sense of help to remove unsafe behaviours among peers.
- Human error can fail all safety systems. Safe behaviour and safety systems must go together. Observers are parents of BBS.
- BBS means back-bone of safety. It is a brotherly approach not command. It is fatherly to motherly, dogly to godly behaviour and its pleasure not pressure. Its based on request not order.
- "*I will sweep away all unsafe behaviours from my workplace/organization,*" a sweeper.
- BBS is a collective call for safe behaviour in the organization.
- "*Yesterday I missed but today I did not miss that unsafe behaviour of my colleague,*" a worker.
- *"BBS is a practice which would improve quality of life,"* a deputy general manager.
- The employees who express good deal of understanding of the concept, process and method BBS during their training shall be identified as observers.
- BBS provides leadership skill for safety to employees but there is no scope for blame on management for any failures rather partnering the responsibility for plant safety.
- *"I came to know through BBS that safety is easy to implement by involving all,"* a safety officer from an oil & gas company.
- BBS experience in organizations takes employees to four stages of behaviour: from conscious unsafe and sub-conscious unsafe to conscious safe and sub-conscious safe behaviours through regular observation and correction process.

- BBS also drives organizations from reactive to dependent to independent to interdependent modes for safe behaviour.
- People need to be greeted when they behave safe.
- "*People have fundamental right to go home safely without a band-aid if they came to workplace without a band-aid*", Mr. Bindra, Director, L&T.
- "*BBS is a business need today as a single unsafe behaviour leads to business slow down*," a senior manager.
- "*We can achieve zero unsafe behaviour by human touch, act and save people around you before you feel guilty, man is for man, unsafe behaviour is also impulsive*," a manager.

5.17 UNSAFE BEHAVIOURS IDENTIFIED BY BBS TRAINED OBSERVERS (IN YEAR 2011)

S. No.	Organization type & no. of employees	Safe behaviours %	At-risk behaviours %	At-risk behaviours corrected by observers %	Average At-risk behaviours per employee %
1	Atomic energy, 1200	61	39	28	4.5
2	Chemicals, 1300	76	24	70	4
3	Automobile, 1000	42	58	63	6
4	Petroleum, 150	54	46	65	2.5
5	LPG, 180	63	37	69	3
6	Petroleum, WI 120	67	33	52	4
7	Petroleum, EI 200	63	37	45	4.5
8	Petroleum, SI 250	83	17	71	4
9	Engineering, 5000	75	25	69	7

10.	Lube Blending Plant, 300	71	29	50	3
	Grand Average	65.5	34.5	58.2	4.25

Our plants in India show **65.5%** safe behaviours and **34.5%** at-risk behaviours. The trained observers not only learn an art and application of BBS, they also learn correcting unsafe/ at-risk behaviours. According to some workmen, BBS is a vaccination to prevent the disease of unsafe behaviour and a trained observer activates all safety systems in an organization. According to a senior manager, "Top-down approach has miserably failed; hence BBS observers are required for plant's safety. Observers must act as visiting observer to other plants to share their experiences."

5.18 BBS PROJECT DELIBERATIONS

During a 3-days training intervention with 257 participants at Lube Blending Plant Chennai India including top/ Senior/middle managers, operators, truck drivers, house-keeping staff.

1. BBS is an 'art of service' of removing an unsafe behaviour in such a way that an observee feels delighted like someone serves you a hot soup or dessert of your choice.
2. How an observer approaches an observee makes all the difference in the attitude of an observee?
3. First line managers have taken BBS activities more seriously than their senior counterparts.
4. Quick intervention is needed on the part of an observer to correct an unsafe behaviour as soon as he observes it. Otherwise unsafe behaviour might show an adverse effect on an observee.
5. Monthly incentive to the observers who make minimum 18-20 observations is necessary for their motivation.
6. "*All times, all places, I would ensure safety*", a lady observer.
7. "*I just feel confident about my plant safety after having trained 30 BBS observers*", chief manager.
8. The first visit of observers to the plant is considered a pre-observation exercise when observers establish first

contact with the observees and explain the concept of BBS.

9. Open house session with a large group (150+) helps creating awareness and acceptance of the BBS concept in the organization.
10. The last level of employees such as drivers/house-keeping/contract staff to be made aware of BBS concept as anybody who enters the plant may commit an unsafe behaviour.
11. BBS is a life and humanity perspective as it is not just situation resolving.
12. BBS at home: we learned to save lives and concern for safety of others at home when we were young as parents taught us, we need to continue that learning and its application.
13. Safety does not delay; it enhances production as it reduces accident-related lost time.
14. Negative attitude of management should not impact the observer as when we use BOFP; there is only me and my observee who needs to be saved from an unsafe behaviour. Let not management's fear affect my interaction and objective with an operator.

5.19 HOW TO STEP-UP BBS PROJECT: AN ACTION PLAN DRAWN DURING A BBS REVIEW MEETINGS AT BILAG WITH 75 STAFF/WORKMEN OBSERVERS ON OCTOBER 2011

1. Every observer will make observations and corrections and fill-up checklist once a day.
2. Observers' meeting with steering team/BBS project leaders will take place every 15 days.
3. New observers are to be included and trained as almost 50% existing observers are found to be silent/passive observers.
4. Appreciation of observers by proper incentive and rewards to be decided by the steering team/BBS project leaders. Observer of the month to be appreciated by gift/certificate/display of his name at the main gate.

5. Provision of resources/compliance such as PPE to be monitored by BBS project leaders. Unsafe conditions discussed in BBS steering team must be complied with.
6. External review to be arranged every 3 months till one year.
7. Repeat survey to be conducted to identify changes as compared to previous survey.
8. Project related contract workers to be educated regularly by observers.
9. Revision /simplification of the existing BBS checklist. Checklist only to be tick-marked. Comments, etc., are optional. Filling of checklist not to be forced upon by managers.
10. BBS project to be advertised through banners and posters across the plant.
11. BBS project to be linked with performance appraisal.
12. Quality of observations to be emphasized by all categories of behaviours observed.
13. BBS concept to be clarified to workers/associates again and again to motivate them.
14. Data entry of the checklist to be done by safety department.
15. Change in managers' behaviour is needed as they are negative, forceful and abusive as reported by most workmen observers.
16. Steering team meetings happened every month but not a single meeting with observers.
17. Production targets are emphasized more; hence managers are not serious about BBS.
18. Improvements have seen in use of PPE, housekeeping, use of tools as a result of BBS.
19. Continuous refresher BBS training for all employees is required for change in their attitude for implementing BBS.
20. BBS observers checklist 'drop box' is needed which must be kept in safety department.
21. Writing name of observer on the checklist should be optional as managers point out to observers who did not fill-up the checklist and abuse or behave negative.

22. BBS need to be implemented with six thumb rules i.e. simplicity (not document-driven only), positivity, mutual regard, determination, patience, and regular interactions.
23. Improvement in safety has increased from 40 to 60% within one year of BBS launch. If the above suggestions are taken seriously, the improvements would speed up greatly.
24. The above action plan is drawn through a day-long serious discussion with observers; hence need to be implemented religiously to see a positive next step in BBS project.

5.20 FAMILY PERSPECTIVE OF BBS

It is only in India that the BBS observers give reference of family members when they talk about safety to their observees. Often they say, somebody is waiting back home, so better behave safer. It is interesting that family reference is joined with behavioural safety. This really touches the coworkers and they learn to behave safe whiling doing their work.

5.21 THAT SAFETY DELAYS PRODUCTION IS A MYTH

Safety delays production is a myth as any first aid injury causes a huge production loss which can be avoided through BBS. The injury-related production loss gets reduced by minimizing unsafe behaviour by observers, hence safety actually fosters production. When an injury takes place in a plant, almost all workmen talk to each other, tend to know what, where and whom the injury happened, so at this time, almost entire plant's production gets affected.

5.22 IGNORING UNSAFE BEHAVIOUR MEANS APPROVING IT

Every time an unsafe behaviour is ignored, it is taken as sanctioned and a silent approval by the workmen and also more it is understood as safe behaviour. So ignoring unsafe behaviour of employees would mean to create an unsafe workplace.

5.23 EACH UNSAFE BEHAVIOUR IS A RISK FOR THE ENTIRE PLANT

Each unsafe behaviour is not only a risk to workmen, but also for others around him and also it is risky for the entire organization. In Jaipur, the entire marketing terminal of IOCL was on fire for12 days due to a single unsafe behaviour of an operator.

5.24 TOTAL SAFETY CULTURE

Total Safety Culture (TSC) can be created if we have active safety systems combined with BBS implementation in the plant. Safety systems (such as SOP, work permits, training, incentives, LOTO procedure, inspection, audits, incident analysis, mock drills, celebration, certifications, compliance, etc.) are necessary and prepare the workforce with the safe attitudes, but it does not reflect in their safe behaviour. Hence, attitude-behaviour gap exposes an organization with the at-risk behaviours of employees which is now being tackled by implementing behavioural safety.

5.25 CONCLUSIONS ON BBS

The above reflections have provided many important aspects of BBS such as accidents/injury don't spare anyone, even managers, the vital characteristics of BBS observers, several conceptual extensions of BBS, positive changes attributed to BBS, managerial perspectives and relevant questions on BBS, and finally BBS is a request for safe behaviour. The trained observers reflected 15% increase in safe behaviours within 2 days of BOFP application. "*I did not know a lot of unsafe behaviours, when observers went for 15 minutes to different work areas, they brought out several unsafe behaviours existing at the workplace, there was a behavioural change from resistance to reception among workers due to BBS training*", the terminal manager.

The lessons learnt from the BBS Indian research are that:

a. The series of unsafe behaviours occur ahead of any near-misses, injury, fatalities.

b. Hundreds/thousands unsafe behaviours are noticeable at any workplace on daily basis depending upon the size of an organization.
c. Showing zero accidents record and international certifications do not really ensure safe organization unless we target zero unsafe behaviours at workplaces.
d. BBS interventions have demonstrated fall in unsafe behaviours and rise in safe behaviours. BBS training also assists in reducing the number of unsafe conditions in the organization.
e. The managements have started believing that engineering controls alone do not provide adequate safe workplace unless behavioural safety is practised and unsafe behaviours need to be controlled in order to ensure total safety at workplaces.
f. Though OHSAS 18001:2007 has included three clauses that emphasize behavioural aspects of safety, the organizations have yet not followed it exactly as the OHSAS 18001:2007 does not provide any guidelines on how to implement these clauses.
g. The Indian multinational organizations have begun to consider the human behaviour aspects of workplace safety more as compared to yesteryears.

"*BBS is the best policy to run my plant safely, and our target is to achieve zero unsafe behaviour rather zero accident*" – a manager.

6

BBS Implementation in a Large Engineering Company

Behaviour Based Safety (BBS) has evolved across cultures with a bit of difference/ twist in its theoretical framework, research and practice and appreciation for organizational requirements.

The Indian mentality is that we would live long as per our fate and destiny irrespective of whatever unsafe behaviours we do. Thus, we Indians behave as '*khatron ke khiladi*' (playing with risks), instead we need to be '*suraksha ke sainik*' (safety soldier).

Safety culture and safety behaviours are two major concerns of top managements in Indian organizations as many people die or get injures as result of workplace accidents.

Safety in documents and implementation are two separate functions or behaviours. It is necessary that what is reflected in safety documents is implemented in behaviours of employees.

The team leaders, unit heads, managers, supervisor, union leaders, contractors, workmen, etc., were trained (during Jan.-April, 2012 in Gujarat state) in the concept, measurement, practical, analysis, implementation of behaviour based safety (BBS) in group size of 25-50.

"*Reason why I live safe is for my grandchildren,*" Mr Bindra, Director L&T.

Vice president (EHS) remarked "*If safety has to become a culture, it has to change in behaviour, we are not doing something which is sustainable, and how many of us reach out to contract workmen, behavioural safety is sustainable. Today, institutionalization of safety is poor; today safety makes a lot of business sense. Each of us can make a culture of safety.*"

DuPont suggested through safety perception survey that we have all safety systems and procedures in place but its implementation in behaviour of employees is lacking. We lost seven lives in the past two years and had 60 reportable accidents. Now we have launched BBS for all units covering all employees and contract workmen (EHS, LnT).

The 120 salient features of this BBS project emerged are as follows:

1. **Seven phases of change process in safety culture** through BBS in large organizations:
   a. Change Awareness - about BBS to all
   b. Change Resistance - from some
   c. Change Negotiation - with them
   d. Change Implementation in units
   e. Change Revision - through feedback
   f. Change Stabilization - with continued effort
   g. Change Review - of behavioural trends
2. Organizational change for BBS process for some units may be a slow implementation.
3. Observers need an identity sticker on their helmet for their recognition for BOFP.
4. Sharing positive examples of achievements for BBS implementation from one unit to another would support.
5. Let each unit set its own small goals to implement without much pressure.
6. Selection of BBS trainers can be identified by characteristics such as communication skills, active interest in safety, already taken awareness training about BBS, ready to implement BBS in their unit.
7. **Contractors and BBS:**
   a. We can make BBS implementation truly effective only if we train all workmen and contract owners/supervisors/engineers from all units/shops.
   b. Each contractor having trained in BBS creates awareness among workmen. Each safety Marshall connected with contractors prepares monthly BBS progress report.
   c. Instead of filling up BBS checklist, contractors maintain a BBS diary to enter their daily observations and corrections of unsafe behaviours on their site.

d. The greatest benefit for contractors in implementing BBS (as they observe and correct unsafe behaviours which trigger accidents /safety violations) is that there would be a possibility of no penalty and there is no threat of withdrawal of gate pass for contract workmen.
e. "*Safety means not to experience any harm on the body so that we can go home safely*," a contractual workman.
f. In BBS management, we have to first train all personnel concerned with contracts as 70-80% injuries and accidents happen to them.
g. "*I can help others for their safety, I can train them*," a BBS trained contract workman.

Observer's personality:

- The observer asked the observee, "if I behave unsafe, you must rectify me also." The observee felt good that the observer is showing concern for his safety.
- Body language of an observer and the positive emotions with which he deals with an observee are important for an effective interaction between them.
- Almost every trained observer is contributing 2% of risk reduction through observation and correction of unsafe behaviours.
- Every trained observer contributed 2% risk reduction through observation and correction.
- Observation power has a lot of effect on others, generally people don't engage in unsafe behaviour as they perceive BBS observers around.
- Observers effectiveness can be understood through his behaviours such as:
  - ❑ Being relation focused
  - ❑ Convincing
  - ❑ Assertive
  - ❑ Inspiring
  - ❑ Being result oriented
  - ❑ Correlating past experience
  - ❑ Ensuring safe behaviour
  - ❑ Not ignoring unsafe behaviours
  - ❑ Smiling

- ❑ Giving personal touch
- ❑ Educating procedure
- ❑ Identify 10% observers from those persons who have undergone BBS awareness training for regular observation.
- ❑ Observers can use both hard and soft copies of BBS checklist.
- ❑ Trained observers test each others' observations for quality check.
- ❑ All observers agreed to correct their own unsafe behaviours first and then others.
- ❑ *Observers have to ensure that the change is observable from unsafe to safe behaviour.*
- ❑ The quality of observer is reflected in terms of being respectful to other employees, relation oriented, polite, closing unsafe behaviour, providing human touch, requesting and thanking for safe behaviour, solution focused and being frank, friendly and straight forward.

8. **Criteria to behave with an observee**
   - Requesting
   - Human touch
   - Thanking
   - Safer Solution
   - Friendly
   - Polite
   - Saying consequences of unsafe behaviour
   - Fearless about interaction with new observee
   - Safety for the unknown
9. **Four steps for BBS implementation by observers**
   - Toolbox talk
   - Daily observation tour (shop visit)
   - Displaying banners
   - Reporting near miss/unsafe behaviours
10. **The big question** is how would BBS boost or fit into other safety systems, such as, organizational safety structure, safety observations, incident analysis, performance standards, and contract safety management in terms of bottom-up involvement, making safety a line function, and a positive/ proactive approach.

11. "*It's easier to teach / train BBS to workmen rather supervisors or seniors as they think that they already know or they are not open to new learning*," a lady observer.
12. There is a difference of a fraction of a second between near-miss and fatal accident.
13. Brothers of safety at L&T remind you of safe behaviour at every moment.
14. BBS is the biggest social work and religious activity for saving lives of friends at work.
15. BBS is saving life of any body, any time, any place.
16. BBS is life style and human nature.
17. Be some ones' buddy for safety (BBS).
18. One can apply BBS beyond one's own site and workmen.
19. Unsafe behaviour leads to loss of safety, production and incentive.
20. People take healthy diets, go for exercising but still die due to stress related heart attacks. Similarly, organizations care for all safety systems but people still kill or injure them as they engage in unsafe behaviours on daily basis.
21. BBS is affection based not compulsion based.
22. BBS taps an inherent ability of people to save others as we naturally educate children not to engage in any unsafe activity and extend our hand to a falling person.
23. How big is an accident, how small is an unsafe behaviour? realized a workman.
24. *Khatron ke khiladi se suraksha ke shainik banana hai.*
25. We have all safety procedures, what we need is to have safe behaviour among employees.
26. *Unsafe behaviour leads to safety loss, production loss, incentive loss and also business loss* as it delays delivery to the ordering party. Once it happened, a 55 ton job has fallen due to unsafe material handing behaviour of an employee and it delayed for a month as it again was taken for machining etc.
27. BBS is a 3-win situation, for an observee, an observer and the organization.
28. We would not take calculated risk, rather plan before time and complete it on time.

29. PPE compliance is low during summer time, 2nd/3rd shift, also among 50% employees having more than 20 years experience.
30. Compliance is more when dealing with hazardous jobs especially handling material during night.
31. We would report unsafe behaviours with nine categories under near-miss procedure.
32. BBS is relation-based safety.
33. Zero unsafe behaviour will save me, others and the organization.
34. Alert today, alive tomorrow.
35. BBS is build-bond-sensitize.
36. Initiating BBS would not mean that our safety systems have failed rather this is the only system that involves 100% of employees/ workmen in identifying /correcting unsafe behaviour on the spot and focus on zero unsafe behaviours which is the root cause of all accidents, near-misses and injury.
37. Safety systems exist on paper but safety lacks in behaviour of employees.
38. Brothers of safety remind each other of safe behaviour at every moment.
39. BBS is bottom-up ownership of organizational safety.
40. Unsafe behaviour conducted even once may kill or injure, but safe behaviour again and again is safe.
41. Normally, people extend help to victims of accidents after they met with accidents, but during BBS; people extend help before people meet with any injury or near-miss.
42. For an effective implementation of BBS, the first line supervisors, shop in-charges must be imparted training first as they pressurize workmen to bypass safety.
43. First there was *manthan* and then kaizen projects, and now BBS in our organization. What's this change every time? This is due to changing focus of new safety philosophies, first we focused on safety systems, then on near-misses and on behaviour which is the root cause of all accidents.

44. *Unsafe behaviours such as horseplay, not maintaining housekeeping and using mobile phones while working among young fresh engineering graduate employees, are very common.*
45. Incidents reporting have increased as a result of BBS awareness.
46. BBS is mentoring not torturing people.
47. It may be nice to alert other person by a whistle when observed doing unsafe behaviour.
48. Identify BBS trainers who have gone through awareness training and also making observation rounds.
49. Do we value human life, can we overcome 100 per cent incidents; both these questions are interwoven.
50. Who does a workman first listen to? It is either the supervisor or his fellow workman; both these people need to be trained first in BBS for effective implementation.
51. Unsafe behaviour is like a 'snake and ladder' game which adversely affects both production and safety.
52. BBS training is considered an added qualification by observers to observe and correct unsafe behaviours on daily basis.
53. Unsafe behaviour leads to loss of safety, production, related incentive, business and motivation.
54. In India, on-paper safety systems are in abundance but it lacks in employee behaviour.
55. Some workers behave unsafe even while wearing PPE. So observers have to be watchful that people work with caution even with PPE on.
56. An observer remarked that supervisors should not worry to increase production but enhance productivity among workmen.
57. BBS is an exchange of safety for each other as observers of unsafe behaviours.
58. Workmen are real actors, architects and implementers of BBS at the shop floor.
59. BBS is safety for known as well as unknown persons like visitors, clients.

60. One man's care is another man's safety as an observer feels.
61. Let's care for human machine, which can be ensured with a sense of disciplined safety.
62. Behavioural sensitivity towards others safety is needed.
63. Each unsafe behaviour has taken some one's life; it can take another life again, so observers alert you every time.
64. Each unsafe behaviour is a kind of RDX, an atom bomb, or a virus which keeps taking lives.
65. "I could have saved his life but I preferred to look another side, I feel guilty," a workman said.
66. **People involve in unsafe behaviour** (not wearing PPE or bad housekeeping etc.) innocently and unconscientiously. Remind them politely so that they follow safe behaviour so easily.
67. **Progress of BBS** would be measured in terms of number of observers trained, frequency of observations tour being conducted by observers, analysis of monthly data trends of safe and at-risk behaviours in each unit, number of meetings being held with observers of BBS/EHS steering teams, and display of BBS progress charts as well as BBS banners in the unit shops.
68. So far **we limited safety with rules and laws** but could not find much success, so we are trying full safety in true sense with BBS approach.
69. **If somebody junior is requesting** you for changing your unsafe behaviour to safe one, don't consider it a loss of self pride or respect.
70. "**Others angry reaction does not prevent me** from reminding and correcting them for their unsafe behaviour to saving their life," a Safety Marshall.
71. **"I thank others** if they remind me of wearing PPE," a workman.
72. "When I observe others unsafe that time what matters is what I say that moment that **he could live for some more time**," a supervisor.
73. **"I don't want to be a reason for others accident** or death, so better I remind them on the spot if I see them unsafe," electrical design head.

74. **Unhealthy behaviours** (eating gutka chewing, smoking at workplace) must be added to behaviour checklist as it affects safe behaviour.
75. We need to give a **BBS card** to each employee for recording 'daily corrections on the spot' of unsafe behaviours at their site.
76. **Saving others is generally at our sub-conscious level**, BBS training has brought it to conscious level.
77. **MBBS** means 'Mera Bharat Bane Surakshit'.
78. **'Nothing wrong will happen' i**s a myth, when you see anybody doing unsafe behaviour, it needs to be corrected immediately, and every unsafe behaviour has killed somebody elsewhere.
79. "HIRA (hazard identification and risk assessment) training needs to be imparted to all supervisors beside BBS training," a supervisor.
80. "I remind others of their unsafe behaviour with true heart so that nobody gets hurt," a workman.
81. "Many people depend on me so I have to go back safe, and same is true of others, so I remind others," another workman.
82. DuPont intervention is more of "reporting & documentation nature' and less of 'ground level behavioural change among workmen," an EHS officer.
83. Safety in all places, smile in all faces – a BBS slogan by a safety Marshall.
84. BBS is "how to converse with workmen about safety and reduce gap with them," a supervisor.
85. "I will not hesitate to remind others to correct their unsafe behaviours irrespective of what reaction I receive from them," chairperson, safety committee of ship building.
86. Workmen are better positioned to make corrections of unsafe behaviours on the spot as well mentor their colleagues about safety.
87. "BBS means safety for entire world and caring for all living beings, and safe care right from a cup of tea to workplace," a workman.
88. "BBS focuses on an essence of time which means save people on time," assistant manager.

89. "The storekeeper does not give us material on time which means we have to carry out work being unsafe," a workman.
90. BBS is eliminating unsafe behaviour of colleagues in a friendly and self-motivated manner.
91. Workmen are better positioned to make spot corrections of unsafe behaviour than the staff.
92. BBS is to empower workmen for organizational safety.
93. "I was an informer of safety violations before attending BBS training, and ordering people for safety promotion, now I am reminding them for saving their life," a workman.
94. Casual approach to safety = casualty
95. "I wish to join safety department," a workman opined after attending BBS workshop.
96. "BBS gives us an understanding of how many persons we are making safe each day," an engineer.
97. "I realized that I can save others' lives, I can help others," a workman.
98. "You are nothing but a human being, your designation comes after, save human being, save others, they save you," a workman.
99. It's time that we care for people who work with us; otherwise we keep losing them in workplace accidents.
100. "BBS adds to my good karma, and I have something to return to society by way of saving human lives," an engineer.
101. "Now since for a few days, corrections of unsafe behaviours are going on by BBS trained observers, people are attentive and visibly working with goggles and gloves," a supervisor.
102. Care for peer is BBS. Its love and compassion. BBS is back-to-back safety.
103. BBS is "never turn your head away when see an unsafe behaviour," an engineer.
104. "Please care for safety of a person sitting next to you, if you want him/her to be alive next second. If we don't care then who, not now then when, not here then where," a workman.
105. "A person's life is more important than company's objective," another workman.

106. Like we have electrical maintenance, BBS is safety maintenance.
107. Once Surdasji was going to temple, a person asked, "can you see the god in temple as you are blind," he replied," "I can't see but god can see me." Similarly, a person who is doing unsafe behaviour can't discern but an observer around him can see and save him.
108. BBS means living with safe behaviour.
109. It is important to study one's own unsafe behaviours at home, on road, at workplace:

	At home	**On the road**	**At workplace**
Unsafe behaviours	1.	1.	1.
	2.	2.	2.
	3.	3.	3.

110. It is further important to count / write as below - how many unsafe behaviours I observed and corrected at workplace everyday/month as a trained BBS observer:

Unsafe behaviours	**Observed**	**Corrected**
1. Use of PPE		
2. Housekeeping		
3. Tools & equipment		
4. Body positioning		
5. Material handling		
6. Communication		
7. Safety procedures		
8. Visual focusing		
9. Use of mobile		

111. **Behavioural trends**

S. no	Date	% of Safe behaviour	% of At-risk behaviour	% of At-risk behaviour corrected	Units /dept. / group	Number of observers	Type of observer
1	23 Jan	60	40	61	Sr mgt.	37	Unit heads
2	24 Jan	59	41	70	HES	36	Managers
3	25 Jan	65	35	55	HES	32	Shop-Incharge
4	30 Jan	67	33	63	HES	23	Supervisors
5	31 Jan	-	-	-	HES	43	Workmen
6	1 Feb	63	37	64	HES	25	Supervisors
7	2 Feb	65	35	44	HES	25	Supervisors
8	6 Feb	63	37	58	Boilers	26	Supervisors
9	7 Feb	72	28	33	Boilers	32	Workmen
10	8 Feb	67	33	52	Boilers	33	Workmen
11	9 Feb	79	21	82	Boilers	41	Workmen
12	13 Feb	81	19	68	Turbine	50	Workmen
13	13 Feb	49	51	54	Turbine	34	Staff
14	14 Feb	76	24	87	Turbine	42	Workmen
15	14 Feb	63	37	58	Turbine	28	Staff
16	15 Feb	82	18	82	Turbine	51	Workmen
17	15 Feb	78	22	90	Turbine	50	Workmen
18	16 Feb	62	38	66	Turbine	21	Staff

Contd...

19	20 Feb	50	50	50	Shipbldg.	21	Staff
20	20 Feb	65	35	33	Shipbldg.	19	Supervisors
21	21 Feb	48	52	40	Shipbldg.	17	Staff
22	21 Feb	73	37	50	Shipbldg.	17	Supervisors
23	22 Feb	53	47	51	Shipbldg.	22	Supervisors
24	22 Feb	68	32	72	Shipbldg.	17	Supervisors
25	23 Feb	60	40	91	Shipbldg.	22	Staff
26	12 March	50	50	48	Piping	24	Staff
27	12 March	85	15	90	Piping	30	Workmen
28	13 March	64	36	44	Piping	19	Staff
29	13 March	80	20	85	Piping	35	Workmen
30	14 March	66	34	75	Piping	18	Staff
31	14 March	84	16	87	Piping	28	Workmen
32	19 March	68	32	67	Piping	35	Contractor
33	19 March	62	38	80	Forging	26	Staff
34	20 March	74	26	100	Forging	42	Contractor
35	20 March	63	37	84	Forging	25	Staff
36	21 March	74	26	92	Forging	32	Contractor
37	21 March	67	33	63	Forging	50	Staff

Contd...

S. no	Date	% of Safe behaviour	% of At-risk behaviour	% of At-risk behaviour corrected	Units /dept. / group	Number of observers	Type of observer
38	22 March	66	34	77	Forging	39	Staff
39	26 March	49	51	70	Central S	26	Staff
40	26 March	70	30	99	Central S	22	Contract wk
41	27 March	70	30	87	Central S	26	Staff
42	27 March	62	38	78	Central S	52	Contract wk
43	28 March	35	65	81	Heavy ca	21	Staff
44	28 March	59	41	69	Heavy ca	17	Contractors
45	29 March	69	31	77	Central S	24	Staff
46	2 April	75	25	100	Contractors	28	Contractors
47	3 April	83	17	100	Contractors	25	Contractors
48	4 April	68	32	75	HES	27	Staff
49	9 April	73	27	84	HE	12	Staff
50	10 April	68	32	89	HE	14	Staff
51	10 April	77	23	97	HE/LEMF	37	Contractors
52	11 April	72	28	63	HE	15	Staff
53	11 April	59	41	80	HE	09	Contractors
54	12 April	49	51	66	HE	53	Staff

Contd...

55	12 April	71	29	91	HE	43	Contractors
56	16 April	75	25	98	Boilers	43	Workmen
57	16 April	68	32	83	Boilers	44	Workmen
58	17 April	70	30	93	Boilers	44	Workmen
59	17 April	67	33	84	Boilers	45	Workmen
60	18 April	65	35	69	Boilers	31	Workmen
61	18 April	67	33	81	Boilers	55	Workmen
62	19 April	77	23	91	Boilers	38	Workmen
63	23 April	79	21	92	Turbine	52	Workmen
63	23 April	63	37	78	Turbine	07	Staff
64	24 April	71	29	76	Turbine	31	Workmen
64	24 April	85	15	82	Turbine	14	Staff
65	25 April	80	20	91	Turbine	30	Workmen
65	25 April	72	28	70	Turbine	22	Staff
66	26 April	74	26	79	Turbine	64	Contractors
	Average	67%	33%	74%		2138	

Overall average behavioural trends across 10 units for 2138 trained observers over 66 sessions are: safe behaviours 67%, at-risk behaviours 33%, corrections of at-risk behaviours in each observation round 74%. This data reflects that at-risk behaviours exist in work areas and BBS is effective as it has shown positive results in terms of correcting at-risk behaviours.

112. **Personal commitment expressed by the trained observers as below:**
    1. Spreading BBS to other workmen.
    2. Daily observation (*parikrama*).
    3. Closing unsafe behaviour on the spot.
    4. Making posters and visual display of BBS banners.
    5. Translation and sharing of 48 BBS principles in local language.
    6. Tool box talk (TBT) on BBS.
    7. Making BBS a part of monthly EHS checklist.
    8. Induction training of new employees on BBS.
    9. Educating supervisors on BBS principles.
    10. Thank you note to best observers.
    11. To have area-wise, shop-wise, and contractor-wise observers.
    12. Taking safety suggestions from workmen.
    13. Inculcating own safe behaviour as observer.
    14. Would share BBS in department meeting.
    15. Follow-up to close unsafe behaviour.
    16. Monthly BBS meeting.
    17. SOP implementation.
    18. Showing film for contract workmen about unsafe behaviour/ condition in the shop.
    19. Complimenting safe behaviour in public and counselling for unsafe behaviour in private.
    20. Being more alert when everything goes right, that's the time when people are behaviourally less alert
    21. Updating myself in safety procedures.
    22. Putting BBS in PO terms for contractor.
    23. Inviting safety person for audit.
    24. No negative approach for correcting unsafe behaviour.
    25. Motivating observers.

26. Direct contact with contract workmen for identifying and correcting unsafe behaviour.

113. **Train the BBS trainers' programme**

The main aim of this program is to develop a BBS system by involvement of dedicated team of BBS trainers for each unit.

BBS programme title: RIGHT TO SAFETY AT WORKPLACE (RSW)

Features of the programme:

1. The 4-day 'train the trainers program' shall include:
   a. Developing as BBS trainer.
   b. The roles and responsibilities of trainers and BBS teams in each unit.
   c. Observation round/tour of the shop/site.
   d. Creating awareness to everyone in respective units.
   e. Difficulties/ issues in BBS implementation.
   f. Forming a BBS steering team.
   g. Functions of BBS steering team.
   h. Developing the observation data sheet and analysis.
   i. Preparing & exhibiting BBS banners at the unit.
2. Each trainer will have a separate ID for entering data on the centralized system.
3. Each unit will have about 40 trainers who are permanent observers.
4. All units will have a well defined BBS steering team. Members of this team will be the BBS observers as well.
5. Each trainer/observer will make an observation round daily/weekly of his unit.
6. Each unit will have 4 days 'train the trainer program' which means about 40 days for 10 units.
7. At the end of 40 days we shall have achieved:
   a. Awareness to everyone in each unit including contract workmen; and
   b. BBS system development in each unit.
8. Company will issue certificate of attendance and identity sticker to each trainer.
9. The best observer shall be rewarded who makes maximum observation in a month.

10. The monthly BBS meeting shall be chaired by the unit head.
11. The unit that reflects the lowest percentage of at-risk behaviours shall be recognized.
12. Each unit will identify the BBS trainers so that the 'train the trainers program' shall begin by mid of May 2012.

114. **Eight BBS slides for display at each shop is required as mentioned below:**
- BBS is brother to brother safety.
- BBS principle is to observe and correct unsafe behaviour on the spot.
- Thanks for your safe behaviour.
- One man's care is another man's safety.
- Save others. Save yourself.
- Unsafe behaviour once, safe behaviour again and again.
- BBS is an art of safe living.
- Unsafe behaviour leads to loss of safety, production, incentive, business, motivation and reputation of the company.

115. **OBSERVERS (N = 30) OWN BEHAVIOURS ASSESSMENT (in percentages)**

BEHAVIOUR CATEGORIES	No. of Safe Behaviours	No. of At-Risk Behaviours
Use of PPE Using PPE, e.g., eyeglasses, hearing protection, gloves, hard hat.	50	50
Housekeeping Work area maintained appropriately, e.g. trash and scrap picked up, no spills, walkways unobstructed, materials and tools organized.	67	33
Using tools and equipment Using correct tools for the job, using tools properly, and tool in good condition.	76	24

Body positioning / protecting Positioning / protecting body parts, e.g. avoiding line of fire, avoiding pinch points.	56	44
Material handling Body mechanics while lifting, pushing and pulling, use of assist devices.	47	53
Communication Verbal and non-verbal interactions that affect safety.	70	30
Following procedures Obtaining, complying with permits, following SOPs, lockout, tag-out procedures…	36	64
Visual focusing (attentiveness)	96	04
Using mobile while working	07	93
Total =	56	44

Note: It is observed from the above table that observers themselves need corrections in their behaviours.

116. **Training outline for BBS trainers/ implementers/ observers: key steps**

1. **Identify almost 400 BBS trainers/implementers/observers** (minimum 2% of entire workforce almost 20,000) from operational areas in consultation with EHS department in each unit. The number of BBS trainers/implementers/ observers would vary in each unit depending upon its workforce size.
2. **BBS trainers/implementers/observers would undergo 4 days training** in a batch of 30 persons; which means there would be a total of almost 13-14 batches to be trained in almost 50 days.
3. The **functions of each BBS trainers/implementers/ observers** would be to:
   - Impart **awareness training** on BBS to a minimum of 50 persons in his unit during 4 days of his training; hence these trainers would train 400×50= all 20,000 workforce in 50 days time.

- Making **daily observation round** as per BBS checklist in his unit.
- Forming **BBS steering team** in each unit from amongst observers.
- The **functions of BBS steering team** would be CDMA (checklist distribution and collection), data analysis, monthly meeting of observers and advertisement through banners.

4. **Monthly Progress Reviews** (MPR) on behavioural trends minimum for one year with BBS steering teams and observers in each unit would sustain the BBS implementation.

117. **Considerations for future roadmap to institutionalize BBS**

1. BBS implementation requires a strong a well defined/ responsible BBS steering team.
2. BBS project has been attended and approved by all unit heads to take it further down the level so we have all reasons to continue with it.
3. So far about 1200 persons are trained, they may not be filling checklist as we did not formalize the BBS system, and it must not be assumed that there is zero implementation. People surely are making observations and correction of unsafe behaviours in their respective units as all those who attended BBS workshops gave positive feedback both verbal and written at the end of each workshop. We must analyze this feedback to understand mindset of people to impact the BBS system further.
4. There is also a visible change in safe behaviours in different units as reported by people.
5. To continue BBS awareness training for rest of the employees and workmen as an overall average at-risk behaviour is identified as 30% and correction of at-risk behaviour is 60% per observation round.
6. As conscientious management to succeed with BBS, we ought to be continuously engaged with BBS training and process for minimum one year.
7. Remember four steps of BBS implementation:
   - Creating awareness among all – 3 months.

- Formalizing / institutionalizing BBS system – 3 months.
- Assessing differences in safety scenario/statistics – 3 months.
- Sustaining/reviewing/strengthening BBS system – 3 months.

8. To have unit level meeting with unit heads and the concerned people for formalizing BBS at respective units and resolving all implementation issues.
9. To hold monthly meeting with appointed teams to strengthen identified 5 focus areas.
10. Monthly meeting with EHS officers to discuss/address any implementation issues.
11. To appoint unit level BBS coordinator.
12. To link BBS observation with performance objective.
13. To commence train the trainers program at the earliest to speed up awareness of BBS.
14. To get nomination of minimum 10% observers in each unit and provide identity sticker as BBS observer.
15. To display BBS banners in each unit for visibility of the BBS project.
16. As BBS implementer across India, I need to hold a meeting with the special team appointed to look after the future of BBS progress.
17. We all take a vow that we shall lead the BBS project at L&T to confirm upon the 'Right to Safety at Workplace (RSW)' for each employee, each workman, and each visitor to send him safe and uninjured back to his family.

118. **Action Plan for Project: Peer-to-Peer safety (PPS) at L&T Bangalore**

a. The trained mentors of BBS shall provide awareness training to all in their respective areas and achieve within two weeks.
b. Display awareness posters (BBS) to conspicuous locations.
   - BROTHER-TO-BROTHER SAFETY(BBS)
   - OBSERVE AND CORRECT UNSAFE BEHAVIOUR ON THE SPOT

c. Motivation scheme to staff & workmen (every fortnight).
   - BBS competition will be held monthly to motivate workmen's and subcontractors. It will be facilitated with prizes.

d. Monitoring department-wise improvement regularly through team leader.

e. Minimizing unsafe behaviour by implementing check-list and updating for all activities.
   - Pocketbook to be issued.
   - To be done on daily basis

f. Conducting internal meeting monthly and confirm the corrective action done.(EHS Committee meeting date monthly)

g. BBS internal audit will be conducted by Steering Team of observers every three-months.

h. Pep talk, tool box talk and Induction training will include BBS.
   - 15 points to be included in pep talk

i. Checklist format to be displayed in common notice board.

j. Area-wise scheduling of observers on daily basis.

k. Announcement of ceremonial launching to be done.

119. **Conclusion**

The BBS training for more than 1500 people has been perceived as simple and responsive. Now we (the managements) are committed to draw the line of action to implement BBS further as follows:

- To form BBS steering teams at plant level, unit level, and shop level;
- To identify and unit-wise trainers to train the rest of the employees and contractors;
- To identify observers for creating monthly behavioural trends.
- First phase of BBS always creates awareness among people, and then the second phase involves forming a steering team, identifying 'internal trainers' and stabilizing BBS process of observation.

- "The major act of accident prevention in any organization is to having shop rounds, observe and correct unsafe behaviours. The line managers must lead safety, and then only we can really prevent accidents," safety head of turbine unit.
- Due to BBS training, near-miss reporting has increased, now near-miss is better defined, as every near-miss is triggered by several unsafe behaviours.
- "We are world class factory, we need to have world class safety," Turbine unit head.

"We shall relate all safety initiates with BBS," expressed a senior official of the company.

BIBLIOGRAPHY FOR FURTHER READING

1. Ahlgren, A. Jari T. and Oja M. (1983). "Personal Safety Equipment and rescuing in disabling occupational accidents". Journal of occupational accidents, 5(1), 9-16.
2. Algera, J. (1990). Feedback Systems in Organisations, International Review of Industrial and Organisational Psychology, 5, London, John Wiley & Sons.
3. Anton, Thomas J. (1989). Occupational Safety & Health Management, New Delhi: McGraw Hill Book Company.
4. Attock Refinery Team Behavioural Safety at Workplace Published on 18 Jul, 2008 in CSR & Companies, Reports & Surveys.
   http://www.tbl.com.pk/behavioural-safety-at-workplace/
5. Bhagwatwar, P.A. (1979). Psychology of Industrial Behaviour, Mumbai, Sheth publishers.
6. Chaube S.P. (2000). Fundamentals of industrial psychology, Mumbai, Himalaya Publishing House.
7. Chuck B. Pettinger (2001). People-Based Safety: The Optimal Approach to Behaviour-Based Safety. Safety Performance Solutions, Inc. Rockville, MD.
8. Davis, Keith (1983). Human Behaviour at Work: Organizational Behaviour, New Delhi: Tata McGraw Hill Publishing Company.
9. Fenney R. J. (1986). Why is there resistance to wearing protective equipment at work? Possible strategies for

overcoming. Journal of Occupational Accidents, 8(3), 207-213.

10. Gangopadhyay, A and Kaila H. L. (1989). Use of Personal Protective Equipment: Psychosocial Aspects. Report No 21, Mumbai, Central Labour Institute.
11. Geller, E. Scott (2004). Behaviour-based safety: a solution to injury prevention: behaviour-based safety empowers employees and addresses the dynamics of injury prevention. Risk & Insurance, 15(12, 01 Oct) p 66.
12. Gilmer B. Von Haller (1961). Industrial Psychology, 2nd edition, McGraw –Hill. Inc.
13. Industrial Safety Chronicle (2004). Vol. 35, Oct-Dec. Navi Mumbai, National Safety Council.
14. International Conference (1989) on "Strategies for Occupational Accident Prevention" (Special Issue) Journal of Occupational Accidents Vol. 2 Nos. 1-3, June 1990.
15. International Safety, Security & Disaster Management Expo (2005). 18-20 September 2005. Mumbai.
16. Kaila HL. Behaviour based safety in organizations, (2012). Prasad Psychological Corporation. New Delhi, Available **at**: tel. 098107 82203.
17. Kaila HL. **Preventing Accidents Through Behaviour Based Safety.** Social Engineer, Vol.13 (1) 2012.
18. Kaila HL. 101 Reflections on Behavioural Safety - developing injury-free organizations. Industrial Engineering Journal, Vol. 3(1), Jan. 2012, 34-39.
19. Kaila HL. **Preventing Accidents through Behaviour Based Safety**: a research survey summary. Indian Journal of Psychology, Jan. 2012 special issue, 28-36.
20. Kaila HL. Preventing accidents through behavioural safety: a research summary. Jagannath University Journal of Psychology, Vol. 1(1), Dec. 2011, 53-60.
21. Kaila HL. **Preventing Accidents Through Behaviour Based Safety**. Industrial Safety Chronicle, Vol. XLII (3) Oct-Dec 2011, 19-23.
22. Kaila HL. Studies on **Preventing Accidents through Behaviour Safety.** Asian Science, Vol. 6 (1&2) June-Dec 2011, 108-111.

23. Kaila HL. Prevent accidents through behavioural safety. Industrial Safety Review, June 2011, 22-23.
24. Kaila HL. Organizational case studies on behaviour based safety (BBS) in India. Industrial Engineering Journal, Vol. 2(22), April 2011, 27-32.
25. Kaila HL. Prevent accidents through behavioural safety. Engineering Review, May 2011, 415-416. www.engrreview.com
26. Kaila HL. Behaviour based safety in organizations. Industrial Safety Chronicle, Vol 37, Oct-Dec 2006, 83-89.
27. Kaila HL. Behaviour Based Safety in Organizations. Business Manager, Vol 10, No. 4, Oct. 2007, 7-11.
28. Kaila HL. Behaviour Based Safety in Organizations. Indian Journal of Applied Psychology, Vol. 45, April, 2008, 35-44.
29. Kaila HL. Behaviour Based Safety in Organizations, the Urban World, April-June *2008*, Vol. I, No. *5, P. 40-48*.
30. Kaila HL. http://www.authorstream.com/Presentation/kailahl-66592-behaviour-based-safety-organizations-dr-kaila-unsafe-safe-behaviour-observation-feedback-steering-committee-english-presentation-hl-new-education-ppt-powerpoint/
31. Kaila HL. (2008). Behaviour based safety in organizations. New Delhi, IK International Publishing House Pvt Ltd.
32. Kaila H.L., A. Singh, S. Ravishankar,S. V. Kamat, (2002) Essentials of Safety Management, Mumbai, Himalaya Publishing House.
33. Kaila, H.L. Dr.S.Ravishankar, Dr. R.K. Mishra; (2003). Human Factors in Organisational Mangement, Mumbai, Himalaya Publishing House.
34. Kaila HL. Behaviour-based Safety Programs Improve Worker Safety in India. Ergonomics in Design' vol. 18 (4), Fall 2010, 17-22. CA, USA.
35. Kaila HL. Organizational cases on behavioural safety management in India, *Injury Prevention*, 2010; 16:A1-A2 doi:10.1136/ip.2010.029215.4. http://injuryprevention.bmj.com/content/16/Suppl_1/A1.4.abstract

36. Kaila HL. Behaviour Based Safety (BBS): A Case Study Psyber News, March, 2011, Vol. 2(2), 2011.
37. Kaila HL. Organizational cases on behaviour based safety management. Business Manager, vol. 13(9), March 2011, 53-57.
38. Kaila HL. Organizational case studies on behaviour based safety (BBS) in India. Journal of psychological researches, vol. 54(1&2), Jan. & August 2010, 19-26.
39. Kaila HL. Organizational cases on behaviour-based safety management in India. Journal of Psychosocial Research, Editorial. Vol. 3(2) July-Dec. 2008, 1-3.
40. Kaila HL. The bottom-up approach to Safety. Indian Oil News. Dec. 2010, 29.
41. Kaila HL. Behaviour-based safety in organizations. Indian Journal of Positive Psychology, vol. 1(1-2) Sept-Dec 2010, 66-69.
42. Kaila HL. Behaviour based safety in organizations. Indian J Occup Environ Med 2006; 10:102-106. http://www.ijoem.com/article.asp?issn=0973-2284;year=2006;volume=10;issue=3;spage=102;epage=106;aulast=Kaila
43. Kaila HL. Behaviour Based safety in organizations. Industrial Safety Chronicle, Vol. 37, Oct-Dec 2006, 83-89.
44. Kaila HL. Behaviour-Based Safety Management: Case Studies across Industries in India, The Indian Journal of Social work, Volume 70, Issue 3, July 2009, 489-501.
45. Kaila HL. "Behaviour based safety in organizations," Business Manager, Vol. 10, No. 4, Oct. 2007, 7-11.
46. Kaila HL. Behaviour based safety in organizations. Indian. Journal of Applied Psychology, Vol. 45, April, 2008, 35-44.
47. Kaila HL. Behaviour based safety in organizations. Urban World, April-June 2008, Vol. I, No. 5, P. 40-48.
48. Kaila HL. BBS winning over employees in India. Occupational Health & Safety, December 2008. *http://ohsonline.com/Articles/2008/12/BBS-Winning-Over-Employees-in-India.aspx*
49. Klen T. and Vayrynen, S. (1984). "The Role of personal protection in prevention of accidental injuries in logging work". Journal of Occupational Accidents, 6 (4), 263-275.

50. Krause, T. (1995). Employee Driven Systems for Safe Behaviour, Van Nostrand Reinhold, New York.
51. Krause, T.R. (1990). The Behaviour-Based Safety Process, New York: Van Nostrand Reinhold.
52. Locke, Edwin A, Shaw, Karyll N, Saari, Lise M, Latham, Gary P. (1981), Goal-Setting and Task Performance. Psychological Bulletin, Vol 90(1), Jul 1981, 125-152.
53. McSween, T.E. The Values-Based Safety Process: Improving Your Safety Culture with a Behavioural Approach. Wiley & Sons, New York, 1995.
54. Marsh, T. et al. (1998). The role of management commitment in determining the success of a behavioural intervention. Journal of the Institute of Occupational Safety and Health, 2 (4).
55. Matthews, Grainne A. Behavioural safety from the consumer's perspective: determining who really provides behaviour safety. Cambridge Center for Behavioural Studies. <http://www.behaviour.org/safety/consumer.cfm>
56. Moran J. B. (1987). A review of the recent criteria of NIOSH for the protection of the work, environment. International Congress on Safety. Health and Environment, Bombay: National Safety Council.
57. Morgan, W. P. (1983). Psychological problems associated with wearing of industrial respirators: A review. Am. Ind. Hyg. Assoc. J., 1983; 44: 671-677.
58. Mosteller, W.G. (1989). Usability Analysis of Messages from a Security System, in Proceedings of the Human Factors Society 33rd Annual Meeting, 399-403. Austin, Texas.
59. OHSAS 18001 Health & Safety Standard. http://www.ohsas-18001-occupational-health-and-safety.com/
60. Pearse, A. (1997). How a behavioural approach helps get to the root of the problem, p. 20. Cited in Managing the Risks of Organizational Accidents, ed. J. Reason, United Kingdom: Ashgate Publishing.
61. Reason, J. (1990). Human Error, Cambridge University Press.

62. Reason, J. (1997). Managing the Risks of Organisational Accidents, Ashgate Publishing.
63. Ridley, J. (1983). Safety at Work. London: Butterworths.
64. Saxena, S. K. and Kaila H. L. (1990). Design of Safety Training Package for a Heavy Engineering Organization: Research Report. Bombay: Central Labour Institute.
65. Schultz Duane, (1973). Psychology and Industry Today. Macmillan Publishing Co. Inc.
66. Skinner, B. (1974). About Behaviourism, Jonathan Cape, London.
67. Stewart, M.G. (1991). Dependence of Human Error Probabilities, in Ergonomics and Human Environments. Proceedings of the 27th Annual Conference of the Ergonomics Society of Australia, Coolum, Australia.
68. Weiss, W.H. (1988). Supervisor's Standard Reference Handbook, New Delhi: Prentice Hall of India Pvt. Ltd.

7

Emerging Issues and Outcomes of Behaviour-Based Safety (BBS) Implementation

7.1 INTRODUCTION

Employees violate safety rules inside the company as children do in school. Employees begin compromising safety rules first by not wearing PPE (personal protective equipment) and gradually go up to violating the SOP (standard or safe operating procedures). These unsafe/ at-risk behaviours on part of employees commence the journey for injuries and fatalities (Kaila, 2012; 2014). According to security personnel of a company, "fifty per cent of contract workers inside the plant are not wearing PPE; we could not enforce it at the company gate, as head of departments (HoDs) would call us to allow them inside the plant without PPE as their work would suffer. On the other hand, the contract workers did not wear PPE as they were not provided with by their contractors and the safety department did not enforce penalty for the same on the contractors. This clearly reflects that they all joined to compromise with unsafe behaviours." With the change of ownership from top to bottom level, safety culture in organizations has dramatically changed (David Broadbent, 2014). According to Fred R. (2014), new BBS programmes evolve through continual training. This requires communication skills among all employees, not just those on the work floor. The management team must be as excited to see progress, and participate with the behavioural observers on a continued cycle, where praise and reward goes a long way towards sustaining the BBS program.

BBS means more people would voice for safe behaviours as being trained observers (Kaila, 2013). According to a deputy general manager of the company, "safety means the number of heads entered in the company premises should be equal to number of heads out of the company gate (with a smile) everyday."

7.2 RESEARCH METHOD

A large gas company in India invited the BBS experts to implement BBS in their gas processing units, petrochemical unit as well as pipelines for almost 4000 employees and 2000 contractors' workmen. The methodology used for implementing BBS is as follows:

1. Conducted BBS awareness workshops for 20% employees, i.e., 800 observers.
2. Conducted Five Days BBS Lead Trainers Workshop for 5%, i.e., 200 employees.
3. Conducted Five Days BBS Roll-out Activities including support and hand-holding with lead trainers' observation and feedback skills.
4. Formed and conducted meetings of Multidisciplinary BBS Functional Safety Committees for Safety excellence and implementation (such as Process, Physical, Maintenance, Road).
    - Process Safety Committee: Review and updates on Safe Operating Procedures, Work Permit System & Lock-out, Tag-out procedure, Management of Change, Training needs, Incident Investigations, Audits, Interlocks, Design Verses operating Parameters, Loading and unloading etc.
    - Physical Safety: PPE need and compliance, housekeeping, ergonomics, equipment and tools, mobile phone usage, rules, regulation and standards on safety and its compliance, etc.
    - Maintenance Practices (mechanical, electrical and instrumentation): calibrations, material handling, electrical tools, energisation and de-energisation, etc.
    - Road Safety: The road safety subcommittee is responsible for assisting the Site Steering Committee

to ensure road safety by developing road safety guideline, travel guideline, rules and regulations for safe driving and implementing other road safety measures, etc.

This is an on-going study and data in this article dates between September 2013 and May 2014.

7.3 DEALING WITH ISSUES AND APPREHENSIONS FACED BY THE BBS LEAD TRAINERS/OBSERVERS

Multidisciplinary lead trainers/observers from various units/areas faced various issues and apprehensions while implementing BBS and these are dealt with as below:

a. How to deal with stubborn employees for correcting their unsafe behaviours? – By persuasion, follow-up, coaching or challenging them.
b. How to deal with senior employees for correcting their unsafe behaviours? – It is a matter of fear versus saving life. One should not make discrimination between junior/ senior employees in saving human life while dealing with at-risk behaviour which could trigger serious consequences such as death.
c. People could enter fake data in BBS checklist of the safe/ unsafe behaviours observed. Restrict filling-up of BBS checklist by trained observers and lead trainers. Random check on the quality of observation by lead trainers. Compare percentage of behavioural trends with ground reality of corrected behaviours on the shop floor. Being over-critical/ over-anxious about future implementation to be avoided. There is no significant trend of fake data reported from other organizations. Hence, better focus on observation process outcomes and discuss regularly. Contract workmen may or may not fill observation checklist. They are basically trained to observe and correct unsafe behaviours around them.
d. How to reward a best observer? Best observer could be rewarded for maximum number of observations as well

as corrections of unsafe behaviours, maximum number of observations days, and maximum number of behaviour categories observed.

e. Whether focus on observation or measurement: first focus on firming up 'the process of observation and correction' by creating BBS awareness among all employees/workmen. Simultaneously and gradually focus on efficacy and accuracy of measurements of BBS data.

f. Some employees do not involve or connect with BBS activity due to organizational behaviour reasons such as lack of promotion, low job satisfaction, poor interpersonal relations. To resolve this, first sensitize these people to the concept of BBS and then involve them in leadership role for BBS project activities for promoting their self-regard which has been damaged due to negative organizational experiences.

g. In addition to the above, some HoDs/OIC's apprehension for BBS implementation came up in terms of the extent of involvement of people and time due to lack of standard manning.

7.4 EMERGING OUTCOMES OF BBS IMPLEMENTATION

It is a desire of meeting targets that people exercise power and authority, hurry up and bypass safety, and in this process that the poor workmen are disabled or killed. Therefore, we need to challenge our own behaviour first and question our consciousness that is it proper to kill people to achieve targets. This was implicitly confessed by the training participants.

BBS implementation in organizations has positively reflected in audit reports. Safety auditors have appreciated involvement of employees in safety behaviours (Kaila, 2013). An executive director said, "With BBS implementation, everybody would become behaviour safety officer 24×7, 365 days". An engineer in-charge expressed that applying BBS inside the plant depends upon my boss. Therefore the role of unit heads is to encourage people through e-mail messages or personal interaction with observers is necessary.

Each awareness workshop was inaugurated by different general managers each day pumping energy into participants. This is something special. Training and safety departments coordinated very well. Work permit violations have been reported to a great extent. PPE non-use and its lack of provision is a big issue. An acceptance of BBS concepts is reflected through one-day sessions' participants. However, the resistance was also shown by some participants mentioning organizational constraints such as target pressures, lack of standard manning. The site BBS steering team and functional committees are formed. An issue of lowest quotation came up very significantly as the concerned contractors were least bothered or educated about safety issues or providing basic PPE to their workmen. Another issue of production priority over safety required organizational leadership intervention.

The visible outcomes that emerged as a result of BBS Implementation are the following:

1. Developed 359 BBS observers and 56 lead trainers to carry forward the safety culture movement to the remaining staff/contractors. Their enthusiasm and openness is witnessed.
2. Five-day workshops upgraded the lead trainers' clarity of BBS concepts and skill of training delivery to their colleagues. As a result, the lead trainers conducted three supervised workshops with full confidence, faith and conviction in BBS at each location.
3. Lead trainers felt a sense of satisfaction by correcting serious unsafe behaviours by regular reminders. They prepared the BBS awareness training module with much creativity and field examples.
4. Prepared IT enabled BBS Checklist, BBS Observation Card, Banners and Sticker for observers' identity.
5. Average base level of behaviours at sites was found to be 73% safe behaviours, 27% unsafe / at-risk behaviours. As a result of corrections of unsafe behaviours, the safe behaviours have gone up from 73% to 95% that needs to be sustained by continuing BBS observation rounds.
6. Some of the characteristics of lead trainers were found to be: good voice quality, involvement of participants,

clarity of BBS Objective, dealing with questions effectively, creativity and originality of examples/workshop material, high energy level.

7. The receptivity of the BBS programme by employees, CISF (Central Industrial Security Force), contract workmen at the sites is felt to a great extent.
8. The BBS banners are displayed across the plant.
9. The monthly BBS award scheme has been chalked out for observers based on observation data in 4 categories i.e., executives, non-executives, CISF, and contract workers.
10. The BBS Checklist (bound booklet) was distributed to all trained observers for continuing observations.
11. The personal involvement, administrative support and leadership of OIC, the cooperation and coordination of Head (Fire & Safety and IT), and all HoDs is visible and appreciated for all BBS activities.
12. The regular discussions on the daily developments were held between BBS task force members and experts.
13. The guidance and involvement of faculty members was found to be useful for the site to implement BBS in true spirit and to be an incident-free organization.
14. Despite all strengths of BBS implementation, some employees at the sites expressed resistance, pressures of work for target completion, too many audits going on, degree of time involvement and lack of resources such as PPE, etc., which were dealt with effectively and addressed by the officer in-charge (OIC) and BBS experts.
15. The regular highlighting of the fact at the sites that the BBS implementation is a corporate agenda and personal involvement of the Chairman of Corporate Steering Committee, General Manager – HSE and his Team, Task Force Members helped implementation of BBS.
16. Interacting with individual HODs and clarifying their role and involvement facilitated driving BBS.
17. Behavioural changes are visible as a result of implementation as perceived by the lead trainers.
18. Lead trainers perceived that on an average, 75% of safety is exercised by the work-permit issuer, 50% by the receiver/

accepter, 30% by the executors and these gaps are filled by the BBS observers.

19. The closing meetings of the Site BBS Steering Committee expressed satisfaction and looked forward to the first follow-up meeting with BBS experts.

7.5 RECOMMENDATIONS

Unsafe behaviour of anybody in the site/plant is a continuous risk for all persons and the plant till it exists and does not get corrected. Unsafe behaviour howsoever small would become critical for life and property anytime. The following recommendations would help consolidating BBS implementation for maintaining incident-free organization (Kaila, 2013a).

1. Lead trainers/observers must practise observation of one SOP or work permit or Lock-out Tag-out (LOTO) a day besides observation of PPE or housekeeping, etc.
2. It is suggested for the company to set up an e-group on the intranet /website for lead trainers, observers, and those related to BBS from all sites to share their success stories and experiences.
3. Maintaining a 'directory of lead trainers of all sites' at the Corporate HSE and periodic mentoring by the corporate team would facilitate BBS project.
4. The CISF enforcement for 'use of PPE and disallowing mobile' at the gate by employees as well as contract workers is emphasized. The CISF needs to be empowered for the same.
5. Lead trainers are backbone of BBS implementation and their active participation is a must for successful BBS implementation.
6. Lead trainers are also internal change agents within the organization and has to train further observers, weekly review among themselves and sharing observations on fortnightly basis among their departmental observers for continuous improvement in their respective departments.
7. The training honorarium/incentive to lead trainers as per existing company rules for conducting training for

non-executives/contract workmen would be a motivating force.

8. It is suggested for sites to display 'board for best observer' of every month, board for behavioural trends as well as display board for listing names of site lead trainers and area observers.
9. The Corporate BBS Task Force members should simultaneously train themselves and emerge as permanent BBS trainers to sustain this project. Moreover all HSE staff whenever they are on internal audit should also conduct BBS refresher programme for the sites across India.
10. The lead trainers must conduct first BBS awareness session for contractors, their supervisors as well as concerned engineers-in-charge as most safety violations take place from contractors. For example, most of the contractors have not provided basic PPE to their workmen and the company has also not enforced much or put penalty for not providing PPE.
11. Field medical attendant on the site is needed for immediate care for preventing blood loss in case injury. "I have visited the plant area 17 years after joining the company hospital today for BBS observations", a lady medical officer.
12. Completing 100% BBS awareness training by lead trainers across the plant is a challenge of lead trainers. Weekly schedule for BBS awareness training for each lead trainer is needed for spreading awareness across the plant among all employees and contract workmen. This schedule needs to be prepared by safety department along with lead trainers.
13. Validation of observers' training conducted by lead trainers is required which can be achieved by weekly contact/meeting between them for better implementation.
14. Near-miss incident reporting is to be replaced with unsafe behaviours for HSE index as near-miss shall be reduced by corrections of unsafe behaviour itself.

7.6 CONCLUSION

Almost all accidents are a compromise with unsafe behaviours.

Every day, 6300 people die as a result of occupational accidents or work-related diseases, i.e., more than 2.3 million deaths per year (International Labor Organization, 2013). This case research points out that accidents at workplace are caused as a result of compromising with unsafe/at-risk behaviours on part of contractors, engineers-in-charge, security/safety personnel and HODs in order to achieve more production in less time. This fact is agreed by all concerned during deliberations while implementing BBS. This reflects that people are pushed to accidents and killed by organizations in a joint effort of all in any project execution. As unsafe behaviours are promoted jointly, they could be discounted jointly as per BBS approach by including all employees in the process of observation and correction of unsafe behaviours. Almost all accidents can be avoided by BBS implementation accompanied by corporate leadership towards zero tolerance for unsafe behaviours at the workplaces.

These days, workers and executives of Indian organizations are working in a highly competitive and hard-task oriented environment. Sometimes, they have to give the targeted production with less manpower in less time. This results into creation of unsafe conditions due to pressure and they begin adopting shortcuts to safe operating procedures. In such situation, it is required to care for their unsafe behaviours by their coworkers which are missing in Indian organizations.

The safety standards, systems, procedures and manuals are well defined in companies but unsafe behaviours are observable at several execution points. According to an executive director, "we have best safety systems but not the best safety behaviours". The existence of unsafe behaviours precisely reflects the lack of enforcement and compromising with safety standards which is a collective arrangement of HoDs, safety department, the plant head and contractors to meet production targets. At the company gate, if security personnel stop employees who are entering without proper PPE, some senior employee would call up or send email to the security department to allow such people inside the company so that the work does not stop. Such compromise with safety only begins with PPE and then employees and workmen compromise safety standards at every execution level even violating work permit, LOTO, SOP. This is how safety culture deteriorates and accidents happen. Hence, the unsafe or accident behaviours are carried out with the knowledge of all concerned simply to achieve targets.

The unsafe behaviours exist and occur every day at the workplace indicate that the journey for injuries and fatalities for employees have commenced. It is widely recognized that the unsafe or at-risk behaviours of employees are the root cause of almost all accidents in organizations. If we try to analyze why people died in workplace accident, we understand that their unsafe behaviours were not observed and/or not corrected. By increasing number of trained BBS observers, we augment our range of observations and address unsafe behaviours of employees and contract workmen on regular basis. To sustain safe behaviours, we need to observe and correct unsafe behaviours again and again by training behaviour safety observers in every area of an organization. Thus, BBS implementation attempts to minimize or zero down injuries and fatalities at the workplace. "If BBS was implemented earlier, it would have saved a few more lives", a senior safety manager said. BBS is to promote safe behaviours partnering everybody in an organization as an individual change orientation leads to organizational change.

Finally, an overall feeling about the BBS implementation experience is found to be encouraging and satisfying for everyone involved in this case of a large Indian gas company which could be a great support for other organizations willing to implement BBS for taking a step forward to achieving zero unsafe behaviours.

REFERENCES

David, Broadbent (2014). Email communication. 3 Jan 2014.

Fred, R. (2014). Email communication. 24 March 2014.

International Labor Organization (2013). Safety and health at work. Retrieved on May 26, 2014.

http:/www.ilo.org/global/topics/safety-and-health-at-work/lang-en/index.htm

Kaila, H.L. (2012).Behaviour based safety in organizations. New Delhi: Prasad Psychological Corporation.

Kaila, H.L. (2013). Review of BBS implementation in Indian organizations. World Focus, 13 (1), 10-16.

Kaila, H.L. (2013a). Is industry really poised for zero accidents? World Focus, 13 (2), 14-19.

Kaila, H.L. (2014). Are we really prepared for zero accident at work? Business Manager, 16 (9), 15-20.

8

Has Industry Really Prepared for Zero Accident? A Review

8.1 INTRODUCTION

Globally, achieving zero accident at the workplace remains one of the major social and economic areas of concern. Every day, 6300 people die as a result of occupational accidents or work-related diseases, i.e., more than 2.3 million deaths per year (International Labour Organization, 2013). Struggle to manage safety situation at workplaces continues. Safety audits are done, awards have been received, and documents are in place. Still organizations wonder what to do to ensure adequate safety of people. An injury hurts the organization because a factory inspector comes and investigates and asks so many embarrassing questions.

The concept of zero-accident (JICOSH, 2013) refers to the total participation campaign into the organization which sounds attractive in a theoretical sense. But practically, zero-accident seems a far-fetched promise of business managements as human safety would not get an absolute preference over profits from production. There is a safety perspective but implementation remains an unlikely endeavour in view of the accident statistics of almost every organization. Workplaces continue killing and disabling people regardless of a variety of existing safety systems, standards or international certifications. Safe workplace is a myth or reality, this paper probes on the issues involved as depicted in Figure 8.1.

Figure 8.1: Theoretical Framework for Zero Accidents (Kaila, 2012)

Zero accidents/injuries at work place requires

↓

Safety (change) management initiatives

↓

Review of safety systems & behavioural safety implementation

↓

Appraisal of organizational safety values
(human life considerations)

↓

Safety leadership (at all levels)

↓

Linking safety to employee performance objectives

Zero accidents/injuries perspective requires a host of changes in management strategies and initiatives for shifting existing safety situation to a new level, which would need a stringent review of the prevalent safety systems and implementation of the latest behavioural safety approach (Matthews, Grainne A. 2013). Further, it is necessary to appraise the organizational safety values especially the core considerations to human life. Above all, the central part is the safety leadership at all levels across organization and linking safety to employee performance objectives is vital.

Though this framework provides a strong theoretical base, practically what we lack is implementation especially in view of organizational executive leadership. According to Health and Safety Executive (2013), effective health and safety performance comes from the top; members of the board have both collective and individual responsibility for health and safety.

Directors and Boards need to examine their own behaviours, both individually and collectively. As per the Head of Environment Health Safety (EHS), L&T, Infra, "The Rules, Code of Practices, Standards and Safety Manual – all these are well defined and readily available as mandatory requirement, but the roadblock is: "*Implementation during job execution*". We are achieving only 50% in implementation that too in limited construction activities. There is cascade effect. It's a paradox that "no cost for safety".

There must be proper budgeted cost for safety since the conceive stage of any project. Through a cut-throat competition one bagged a contract; he always endeavours how cost effectively the job can be executed. Obviously "*safety takes back seat*". It's a long way travel to go before we say "*Indian construction industry is poised for incident and injury free*" (Personal Communication, 2013).

According to Japan International Center for Occupational Safety and Health, there are three pillars of the zero-accident campaign: the positive attitude of the top management; the complete management of the safety and health system by line managers and supervisors; and promotion of voluntary activities in the workplace (JICOSH, 2013).

This article is an extract from the qualitative/narrative data of a longitudinal national action research survey on behaviour based safety in India. It's an on-going study in which almost 500 BBS training workshops were conducted for nearly 9300 workmen (operators, technicians, contractors) and union representatives and about 1800 managers at all levels between 1997 and 2013 in diverse multinational organizations across industrial sectors. The Indian companies which participated in such BBS programmes were Maharatna / Navratna companies, energy giant, chemical, automobile, pharmaceutical, electrical, nuclear, steel, construction, shipping, coal, heavy engineering, and so on). They were exposed to the BBS concepts; observation and feedback processes; and implementation of BBS in an organization. The workshops were held in English/Hindi languages of about 30-50 per batch of training participants. It is true that India, as predicted by many, is to become the world's third largest economy by adopting the best work systems such as behavioural safety.

8.2 REASONS FOR NOT ACHIEVING ZERO ACCIDENTS AT WORKPLACE

A senior HSE professional of Indian Oil Corporation points out, "It is only desire or we can say target to have zero accident. But, when we work practically in the field, our concentration is always directed towards more production to meet the market demand.

The most important issue is development of infrastructure which is happening along with the growth of demand. And, this creates pressure on production and safety system is bypassed. However, in recent days, everywhere management as well as workers have become cautious while working in the field".

The best safety plans and programmes don't continue and/or show a failure as a result of inactive leadership behaviours across departments in an organization. Hence, the zero accident target remains a myth and does not convert into reality. The application of behaviour-based safety (BBS) approach is facing a severe challenge that safety performance may decline when BBS intervention is removed (Mingzong Zhang, Dongping Fang, 2013). Zero accident targets appear to be a very difficult proposition for organizations for a series of reasons (Kaila, 2012) as below:

a. Organizations lack in strong implementation of safety systems.
b. Organizational leadership vary in emphasis over human life safety.
c. Business is a clear priority over safety function.
d. Safety is mostly not a performance objective.
e. Every industrial project has killed one or more persons (mostly contract labour) then what would make us think that zero accident is achievable.
f. Zero accident looks perfect in documentation, not in real work life.
g. For most of the organizations, safety is a staff function; employees are not involved as per OHSAS 18001: 2007 clauses on behavioural safety.
h. The Hundreds of unsafe behaviours are noticeable at any workplace on daily basis depending upon size of an organization. Hence, the potential for accidents exist in every organization unless unsafe behaviours are tapped and controlled daily. So it would not be a surprise if there is a minor or major accident any time due to these unsafe behaviours.

i. Showing zero accidents record and international certifications do not really ensure a safe organization unless we target zero unsafe behaviours at workplaces.
j. Organizational case studies revealed that a single unsafe behaviour can prove to be fatal.
k. Lack of or partial safety enforcement reinforces unsafe behaviours at the workplace.

To reduce accidents, the managements have executed safety interventions such as risk assessment, suggestion scheme, training, safety committee, auditing, motivational programmes (quiz, award, incentives), SOPs, plant inspection, work permit system, etc. Most of these safety management systems have aimed at controlling unsafe conditions, whereas most of accidents are triggered by unsafe acts or behaviours, even unsafe conditions are the result of unsafe behaviours.

Zero accident would mean total safety culture which can be created if we have active safety systems combined with BBS implementation in the plant. Safety systems (such as SOP, work permits, training, incentives, Lock-out Tag-out (LOTO) procedure, inspection, audits, incident analysis, mock drills, celebration, certifications, compliance, etc.) are necessary and prepare the workforce with the safe attitudes, but it does not reflect in their safe behaviour. The attitude-behaviour gap exposes an organization with the at-risk behaviours of employees which is now being tackled by implementing behavioural safety.

8.3 TARGETING ZERO UNSAFE BEHAVIOURS TO ACHIEVE ZERO ACCIDENT

Behavioural safety postulates that unsafe behaviours are at the base of any near-misses, injury, fatalities (Figure 8.2). If unsafe behaviours are controlled, even the near-misses can be avoided.

Three clauses of OHSAS 18001:2007 document emphasize behavioural aspects of safety in Occupational Health & Safety (OH&S) Management Systems that organizations shall establish, implement and maintain a procedure(s) for: the participation of workers by their appropriate involvement in hazard identification, risk assessments and determination of controls; for hazard identification and risk assessment that shall take into account:

human behaviour, capabilities and other human factors; and to make persons working under its control aware of the OH&S consequences, actual or potential, of their work activities, their behaviour, and the OH&S benefits of improved personal performance. Behavioural safety popularly known as behaviour based safety (BBS) encourages contribution towards safety from everyone in the organization.

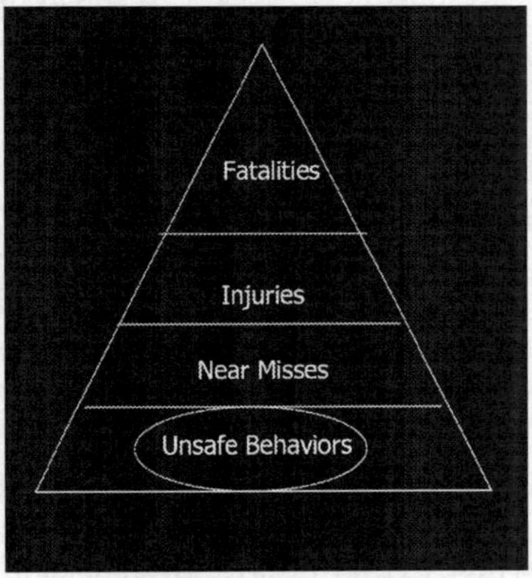

Fig. 8.2 BBS Triangle (Kaila, 2008).

"Due to BBS implementation, the accident frequency rate from April 2012 to May 2013 indicates reducing trend in general. However, we are trying to improve the performance further with the challenges being the high manpower turnover. BBS is the best policy to run my plant safely, and our target is to achieve zero unsafe behaviour rather zero accident," a manager said.

Behavioural trends reveal that at-risk behaviours exist in every workplace which can trigger accident any time. Poor knowledge, short of supervision and maintenance are the obvious reasons for accidents but these factors are actually nothing but the unsafe behaviours of the concerned people for not carrying out their jobs responsibly. The socio-cultural backdrop of employees may also impact the practice of unsafe behaviour (Kaila, 2013).

The unsafe / at-risk behaviour are also caused by some HoDs who emphasize/ force completing the work by avoiding basic safety rules or regulations, e.g., allowing under-capacity cranes/

hydra to lift/shift heavy jobs. The project manager asks the site engineer to complete certain jobs by today itself. He carries out the jobs somehow by-passing safety regulations. He can't negate the project manager and make him unhappy. This is a very common scenario in construction sites. It takes a second of unsafe behaviour and turning into an accident or big fire and delaying the production process for long time.

Unsafe behaviour can happen to anyone regardless of position, education, experience and age. A vice-president went up on the fourth floor to inspect a construction project, he received a call on his mobile and started talking, got so engrossed that he just put his step forward and fell down from the 4th floor and died on the spot. An engineer on the shop floor thought of crossing a conveyor belt while it was stopped, as he crossed, it started working, he got crushed and died. A deputy general manager got a serious eye injury when he was observing a workman without wearing safety goggles and an object flew from the machine and hit him. So accident/injury spares no one, even managers.

According to a manager, "punishment is also a strong sense of caring like a giraffe-mother kicks her baby's legs at birth to make her move, otherwise it's a prey to others in the jungle". Punishment in terms of fine or loss of pay will set a sense of fear and would not let an employee do unsafe behaviour; it seems contrary to BBS positive approach. But then that fear also reduces the unsafe behaviour for some employees to a certain extent which is indirectly positive for safety of people at work.

8.4 MANAGERS' CONCEPT OF BEHAVIOURAL SAFETY

According to Deputy General Manager of Indo Gulf Fertilisers, the company has given top priority to safety, health and environment right from the design stage. To minimise the possibility of accidents, we have developed people to know their equipment intimately and to be consistently safety, health & environment conscious. The line managers are committed to safety and health. Unsafe conditions and acts observed are brought to the attention of the concerned departmental heads for quick remedial action. Safety consciousness of an employee is given due importance in his annual performance appraisal.

"We implemented BBS at Qatar Chemicals Ltd. in 2008. It has given result and we reached to the certain maturity ladder of the BBS process. As time passes people are losing interest now, observations are not done consistently, clean sheet observations are increasing, steering committee members are not getting motivated and hazards are not identified properly, etc. Sustainability is the main issue right now in our process. I need to revamp the process, what to do next, how I can bring back the energy in the process," said a BBS co-ordinator.

"Safety in Indian scenario is considered a fire-fighting approach for emergency situations. Normally, people complete the jobs fast towards meeting a deadline following a series of unsafe behaviours. An unsafe behaviour being practised by coworkers would once for sure turn into an accident some day. In India, safety departments are understaffed and over worked. It is not possible for them to deliver their jobs effectively for safety of employees and contract staff," said a general manager. BBS provides more eyes for safety in the organization. It has three basic principles (i): interaction between observers and observees, (ii) inclusion of every workman/employee in safety efforts, and (iii) intention to provide injury-free environment.

Top management's role is significant in terms of monitoring, motivating, and mentoring those involved in BBS project, failing which the project may collapse or weaken its purpose. There is an interesting positive turn in the Indian safety scenario. The top managements are emphasizing and communicating their safety concerns down the level. What they are saying is that merely providing safety infrastructure in the organization or location/ sites is not enough or adequate for safety of the people. The HoD/ supervisors need to involve all employees at all levels up to the last person who may be least aware of safety (such as contractors' workmen or housekeeping staff). The section heads opined that we are ready to implement BBS approach but we are apprehensive of top leadership to this project. Many systems have come and gone, there is a lack of long commitment. If BBS is top-driven then we have to do it. On the other hand, the top management showed a deep concern for the section heads' varied level of involvement, credibility, acceptability and safety-mindedness for implementation of BBS with full spirit in their respective units.

"Safety is not a priority; it is the core value which has to emanate from highest level of leadership in the organization, combine safety systems with BBS", Director – L&T.

8.5 RESEARCH IMPLICATIONS

"Whether we are in business or not, we need safety. One accident can change our life. We have created little vibration for safety in industry, is it enough? We need to work more on safety and develop a culture in which every employee understands safety. We need to train them so that they have the mindset for safety. Safety is strongly linked to the human thought and behaviour. It's a culture which we would not get in a day. It needs persistence," said Vedanta plant head.

According to a HSE's vice president of a company, "All failures of health, safety, and environment are around behaviours". Behaviour based safety (BBS) approach is a planned effort and an organizational intervention/ procedure for making safety a true line function. Though Occupational Health and Safety Assessment Series (OHSAS) 18001:2007 have included three clauses that emphasize behavioural aspects of safety, organizations have yet not followed it as OHSAS 18001:2007 does not provide any guidelines on how to implement these clauses.

According to an EHS head, "we are expanding our business, growing revenues but every year we have 40-42 fatalities in all independent companies (ICs). We are not factories to kill people or to create permanent disabilities due to accidents in our work premises. We have hardware but not the software which is the 'feeling', this is the 'disconnect'. We have not valued human life. And now as per new guidelines, if there are more than three fatalities at our workplace, we are not qualified for tender registration for any new business. So safety of people is related to business.

The managements have started believing that engineering controls alone do not provide adequate safe workplace and unsafe behaviours need to be controlled in order to ensure total safety at workplaces. The Indian multinational organizations have begun to consider the human behaviour aspects of workplace safety more as compared to yester years. Figure 8.3 below depicts pictures of some BBS programs conducted by the author as part of this study.

According to a general manager (safety), "behaviour-based safety is all about changing the basic organizational culture to inculcate positive safety at the workplace". The body of research literature created through behavioural safety interventions has made us realize the value of macro-ergonomics aspects. It is important to underline that people behave unsafe or take risks even in well designed work stations. People tend to take risks in safe environments and be very alert in unsafe environments. People speed up their vehicles on highways and tend to be very alert in crowded streets. The engineering systems, the process or task design provided at workplaces are the hardware part; and the software part is the behaviours of people who tend to behave safe or unsafe at times. Even with well designed work stations, the workmen/operators tend to engage in at-risk behaviours (such as not using personal protective equipment, work area not maintained appropriately, not using correct tools for the job or tools not in good condition, inappropriate body mechanics while lifting, pushing and pulling material, not complying with work permits or following safe operating procedures (SOPs), and using mobile while working) that may trigger injury or accident. Also several case studies in organizations have revealed that each of these at-risk behaviours has been fatal to workmen, engineers, and managers (Kaila, 2010).

As per the head HSE, ONGC, "Industries are nurturing safety culture these days and focusing on people behaviour as one of the indices for avoiding accident". Behavioural safety is getting well accepted in India and showing good results in terms of improved safety records, building positive SHE (safety, health & environment) culture and reduced at-risk behaviours at workplaces and is also being applied successfully worldwide instead of command-and-control approach to occupational safety (Geller, 2004).

Fig. 8.3: Some BBS programs conducted by author (Kaila, 2012).

Despite best safety programmes being available and exercised, it needs to be understood as to why each year witness hundreds of thousands people being killed at workplaces. Attempting new safety programmes from time to time has not enabled the industry to achieving zero accidents. Does it point out the lack of will power on part of managements to lead and implement these safety plans effectively on regular basis for the safety of human being? The vice president of a giant engineering company Larsen & Toubro expresses, "If you look at Indian Industry as a whole we are not poised for a zero accident scenario. This is because while all safety measures are being taken we still have not mastered the art of 'safe behaviour'. There is not enough visible commitment from senior management of industry to this issue", (Personal Communication, 2013). In the wake of large number of fatalities taking place each year due to workplace accidents, the zero accident targets appear far from reality for industry and we cannot yet claim our workplace to be safe. Journey towards zero accident/injury goes on. The human life value consideration for organizations falls short to the business targets. Human safety is clearly not a core value for industrial world. "In the Indian culture, we are more concerned for life after death than the 'present life'. The 'value for life' is not prominent in the industrial culture due to work pressures", said a project manager.

8.6 CONCLUSION

To wrap up, "*Is our Industry Really Ticked for Zero Accident*", HSE Manager of Jindal Power narrates many things. Any industry is related to physical, chemical and biological hazards. It may not be advisable to say that zero accident means indicating towards injury only. It can be also zero harm. Any industry's safety aspects lie with the HSE policy, adaptation of standard safe practices in organization, commitment, involvement and implementation by top management, empowering safety personnel for putting innovative ideas/techniques in company as per with international safety standards and monitoring of deviations with latest techniques/ software. Again the zero accident indicates about the size of company, the work methodology implemented, deployment of skilled personnel, strict guidelines for adherence to safety norms, etc. Finally, in India many of the industries are prone to accidents as there are no stringent statutory obligations, law abiding factors, no threat for the management, etc. According to a safety practitioner of Bosch, "any activity we do in an industry we face few hazards, be it, improvising on process safety every day. As long as our standards or expectations keep increasing, so does our knowledge on the hazards, we will plan to work without those hazards, and try to look for new ones. But we still are unsure, with all the systems in place, can we be accident free. It's hypothetical until a full-proof-system is created" (Personal Communication, 2013). What really matters is an undeterred active commitment of senior management to implement the safety systems along with their employees and workmen. Managements must aim to achieve business-safety balance for the cause of human safety while emphasizing the behaviour-safety connection in the broader organizational perspective. The capacity building programmes on 'change management interventions' to senior managers in organizations would help to better the safety situation. Safety survives, business thrives.

REFERENCES

Geller, E. Scott (2004). Behaviour-based safety: a solution to injury prevention: behaviour-based safety empowers employees and addresses the dynamics of injury prevention. Risk & Insurance, 15 (12, 01 Oct) p 66.

Health and Safety Executive (2013). Why leadership is important. Retrieved on June 30, 2013. http://www.hse.gov.uk/leadership/whyleadership.htm

International Labour Organization (2013). Safety and health at work. Retrieved on June 30, 2013. http://www.ilo.org/global/topics/safety-and-health-at-work/lang--en/index.htm

JICOSH (2013). Concept of Zero-accident Total Participation Campaign. Retrieved on June 29, 2013. http://www.jniosh.go.jp/icpro/jicosh-old/english/index.html

Kaila HL. (2010), Behaviour-based safety programs improve worker safety in India. Ergonomics in Design' vol. 18 (4), Fall, 17-22. CA, USA.

Kaila HL. (2008), BBS winning over employees in India. Occupational Health & Safety, December 2008. http://www.ohsonline.com/Articles/2008/12/BBS-Winning-Over-Employees-in-India.aspx

Kaila HL. (2011), "Organizational cases on behaviour based safety (BBS) in India". International Journal of Human Resource Management, vol. 22, 10-12, 2135-46, June-July 2011.

Kaila HL. (2012), "100 Reflections on Behavioural Safety: Developing injury-free organizations. Industrial Safety Chronicle, vol. XLIII, 3, 21-26, Oct-Dec 2012.

Kaila HL (2013). Review of BBS implementation in Indian organizations. World Focus, Vol. 13 (1), 1, 10-16. http://viewer.zmags.com/publication/688b5926#/688b5926/1

Matthews, Grainne A. (2013). Behavioural safety from the consumer's perspective: Determining who really provides behaviour safety. Retrieved on 4 July 2013. Cambridge Center for Behavioural Studies. <http://www.behaviour.org/safety/consumer.cfm>

Mingzong Zhang, Dongping Fang (2013). A continuous behaviour-based safety strategy for persistent safety improvement in construction industry. Automation in Construction 34 (2013) 101–107.

OHSAS 18001. (2007). Occupational health and safety assessment series. http://www.scribd.com/doc/10081792/OHSAS-18001-2007-Occupational-Health-and-Safety-Management-Systems-Requirements1. Retrieved on 02 February 2010.

Personal Communication (2013). Email discussion with safety professionals, July 2013.

ACKNOWLEDGEMENTS

I would like to express many thanks to various safety professionals for their contribution to this article, they are: MK Rao of India Glycols, Ranjeet Maity of Qatar Chemicals Ltd, R. Kapoor of Indo Gulf Fertilisers, Suresh K Singh of Indian oil corporation, A.K. Sapre of ONGC, K Ramakrishnan of Larsen & Toubro Limited, Manoranjan Dash of Jindal Power Limited, Subhamoy Maitra of L&T Infra and Smruti Guttal of Bosch Limited.

9

Towards Maintaining Successful BBS Programme

9.1 INTRODUCTION

Organizations introducing new BBS programmes have many questions before they implement the same at their workplaces. The corporate HSE departments seek the involvement of the top management executives, expose them to the concept of BBS to get their consent, and also involve the heads of the sites where they plan to launch BBS. Consequently, there would be a series of pre-implementation exercises and preparations that the HSE departments undertake. Finally, the major task remains to identify the BBS expert team members who would facilitate introducing this organizational intervention with results in terms of desired changes such as positive difference in the organizational safety statistics, i.e., rate and frequency of injuries, accidents, near-misses and at-risk behaviours among employees and workmen.

Some noteworthy opinions expressed by the safety practitioners are as follows:

- "With the change of ownership from top to bottom level, safety culture in organizations has dramatically changed", David Broadbent (email communication, 3rd Jan 2014).
- According to Sri Lankan safety officer, "as long as we keep trying to cut costs at every level, what we really succeed in doing is cutting lives".
- At the inaugural function of BBS implementation on 11 Dec 2013 at Gas Authority of India, the director of marketing operations stated that it is paradoxical to say that we have safety department, in fact everybody should

be a part of safety. BBS is a journey; it's a volitional activity more than compliance.

9.2 RESEARCH METHOD

This chapter is an extract from the data of a longitudinal national action research survey on behaviour based safety in India. It's an on-going study in which almost 600 BBS training workshops were conducted for 10,600 workmen (operators, technicians, contractors, union representatives) and 2400 managers at all levels between 1997 to 2014 in diverse multinational organizations across industrial sectors. The Indian companies which participated in such BBS programmes were Maharatna / Navratna companies, Energy Giant, chemical, automobile, pharmaceutical, electrical, nuclear, steel, construction, shipping, coal, heavy engineering, and so on). They were exposed to the BBS concepts, observation and feedback processes, and implementation of BBS in an organization. The workshops were held in English/Hindi language of about 30-50 per batch of training participants. The participants in BBS programme expected to know the following:

- How to reduce accidents by changing behaviour?
- What is an easily and quickly implementable safety approach for unskilled/casual labour?
- How to teach contract/casual workers for minimizing accidents?
- What is the concept of BBS to improve safety performance?

9.3 UNSAFE ACTS/AT-RISK BEHAVIOURS ASSESSED (DURING JULY 2013 AND JANUARY 2014)

Table 9.1 below reflect that on an average, 84% behaviours of the employees and workmen are perceived as safe and 16% behaviours are perceived as unsafe. This was determined through discussions with HoDs as well as visits to the plant. This data directs us to say that plant has the safety systems /procedures as shown in the nine behaviour categories below but these systems are not reflecting in safe behaviour of the employees and workmen. This confirms the need for BBS implementation for better safety in order to target zero unsafe behaviour in the plant.

It is important to remember here that any single unsafe behaviour is capable of triggering an accident/ injury or fire any time, and if it is not addressed; it has to convert into an incident for sure. An individual category of behaviour below needs to be monitored which must be done during the BBS implementation process.

Table 9.1: Perceived % of Safe Behaviours and At-Risk Behaviours

	Behaviour Categories	Perceived % of Safe Behaviours	Perceived % of At-Risk Behaviours
1.	**PPE** - Using, e.g., eye glasses, hearing protection, gloves, helmet, etc.	85	15
2.	**Housekeeping** - Work area maintained appropriately, e.g., trash and scrap picked up, no spills, walkways unobstructed, materials and tools organized.	84	16
3.	**Using tools and equipment** - Using correct tools for the job, using tools properly, and tool in good condition.	83	17
4.	**Body positioning / protecting** - Positioning / protecting body parts, e.g. avoiding line of fire, avoiding pinch points.	79	21
5.	**Material handling,** e.g., body mechanics while lifting, pushing and pulling, use of assist devices.	78	22

6.	**Communication** - Verbal and non-verbal interaction that affects safety.	80	20
7.	**Following procedures,** e.g., obtaining, complying with permits, following SOPs, lockout, tag-out procedures.	90	10
8.	**Visual focusing (attentiveness)**	83	17
9.	**Using mobile phone** while working	94	06
	Average	84	16

9.4 TRAINING BBS OBSERVERS AND LEAD BBS TRAINERS

According to Fred R. (2014), new BBS programmes evolve through continual training. The employees are like jars of clay. They must be moulded into the model employees that represent the company safety at its highest level. This requires communication skills among all employees, not just those on the workfloor. The management team must be as excited to see progress, and participate with the behavioural observers on a continued cycle, where praise goes a long way towards sustaining the programme.

A broad criterion for selection of training participants as BBS observers and Lead BBS Trainers is as below.

1. Preferably drawn from all sections/depts.
2. Positive minded.
3. Safety oriented/bent of mind.
4. Good communicator, expressive.
5. Ready to create BBS awareness and train people down the level as Lead Trainers as well as support BBS project actively.
6. Ready to make regular observation and correct unsafe behaviours as BBS observer.

The lead trainers are responsible for creating BBS awareness and train people as BBS observers down the level as well as support BBS project actively, and to make regular observation and correct unsafe behaviours. The contents of five-day lead BBS trainers' programme are listed in Table 9.2 for the guidance of organizations.

Table 9.2: Contents of a detailed Five-day Lead BBS Trainers Programme

Agenda: 10 am to 5 pm everyday
Day 1
• Concept of Behaviour Based Safety
• Attitude & Behaviour Change
• BBS as OHSAS Compliance
• Process of BBS
• Observation and Feedback process
• Observer and observee profile
• Observers' eight P behaviours
• Six stages of human change process
• Measurement of safe & unsafe behaviours
• Documentation & display of safe & unsafe behaviours
• Traditional safety vs. BBS approach
• Safety systems and BBS
• Three methods to achieve safe behaviour
• Zero accidents vs. zero unsafe behaviour
• Participants' learning assessment & queries
• Workshop Evaluation & Valediction
Day 2
• Creating BBS trainers inside organization
• Road Map: Implementing behaviour safety across organization
• BBS advertisement: Posters and Banners
• Managements' Questions before BBS launch
• Experiential learning of BBS trained observers
• how to sustain BBS culture in organization
• Observers' test of excellence
• Participants' learning assessment & queries

Day 3

- BBS Project organization
- Formation of BBS Steering Team and its Tasks
- Displaying BBS banners/posters at the location
- Preparation of training material, observation card & stickers for observers
- Participants' learning assessment & queries

Day 4

- Nomination of a dedicated BBS Project Coordinator at the location
- Ensuring strong leadership drive by the OICs/HODs for BBS project
- Linking BBS observation as part of PMS/KRA
- IT enabling BBS Observation checklist
- Developing BBS training as an e-module
- Participants' learning assessment & queries

Day 5

- Ceremonial launch of BBS project by Management at the location
- Announcing BBS project on company portal
- Developing base level of safe& unsafe behaviours at the location
- Issuing an official written announcement of BBS project at the location
- Issuing BBS sticker to the Lead Trainers to be put on the helmet as an identity
- Issuing Training Attendance Certificate to all participants as motivation factor
- Participants' learning assessment & queries
- Workshop Evaluation & Valediction

Note: There is a practical on-site/plant observation visit and analysis every day.

9.5 STEPS OF A SYSTEMATIC BBS PROGRAM DESIGN FOR ORGANIZATIONS

Organizations often wonder for what would go into the detailed BBS implementation programme. Hence, the following steps of a systematic BBS programme design have been adopted (Kaila, 2013a).

1. Formation of corporate level BBS implementation team
2. BBS implementation project kick-off meeting.
3. BBS workshop for Top Management.
4. Announcing BBS project on company portal as corporate initiative.
5. Safety assessment survey at locations.
6. BBS workshop for corporate level BBS Implementation Team.
7. Formation of Steering BBS Team for each location.
8. Nomination of a dedicated BBS Project Coordinator for each location.
9. Conducting BBS lead trainers followed by project guidance at each location.
10. Ensuring strong leadership drive by all HoDs for BBS project activities.
11. Displaying BBS banners/posters at each location for advertisement.
12. IT enabling BBS observation checklist.
13. Developing BBS training as an e-module.
14. BBS Monthly Review by Steering Teams by each location and compliance with its proceedings.
15. Rewarding best BBS observers based on maximum observation.
16. Corporate Apex Implementation Team Review on BBS monthly progress.
17. Including BBS awareness training in contractors / vendors terms & conditions.
18. Linking BBS observations with Performance Management System/Key Result Areas.
19. Safety Statistics Board should also reflect the percentage of SUSA (safe and unsafe acts).

20. HSE Index to include the monthly % of unsafe/at-risk behaviour.
21. Comparing injury data every 6 months before/after launch date of BBS.
22. Celebration of BBS annual day.

9.6 RESPONSIBILITIES OF THE CORPORATE TASK FORCE/LOCAL STEERING COMMITTEE ON BBS

The Corporate Task Force on BBS is an active body to check and oversee that the new BBS programme evolves through continual training, motivation, mentoring and monitoring the progress at each site. The management team must be excited to see progress and participate with the Local Steering Committees on a continued cycle, where praise/recognition to sites which are making headway goes a long way towards sustaining the programme, and regularly pour enriching thoughts towards maintaining a successful BBS programme.

The Corporate Task Force on BBS must ensure the following for successful implementation:

- Safety statistics board at each location should reflect the percentage of safe and unsafe acts.
- Percentage of at-risk behaviour at each location should be a part of HSE Index.
- Safety motivation / recognition need to be introduced at each location for BBS observers.
- BBS observations must be linked to performance objective.

Responsibilities of the Local Steering Committee on BBS at the location/site.

The Local Steering Committee on BBS is an active execution body at each site for actual implementation as well as to oversee and motivate the lead trainers for spreading complete awareness in the plant/site. The local steering committee needs to conduct monthly meeting as per steering team tasks.

9.7 METHODOLOGY AND ACTIVITIES DURING BBS ROLL-OUT

The BBS roll-out across an organization would mean actually supporting, handholding by the BBS expert faculty and ensuring

BBS implementation including awareness programme to all employees/contract workmen by the lead trainers, formation of BBS steering team and clarification of its tasks, BBS banners prepared/displayed, IT Enabling BBS checklist, practical on-site/plant observation visit and its analysis every day (Kaila, 2013b). The roll-out actions are described below in Table 9.3.

Table 9.3: BBS Roll-out Plan of Actions for organizations

Nomination of BBS Project Coordinator
Formal announcement of BBS through written communication by plant head
Formation of Steering BBS
Creating BBS awareness in the staff and contract workmen
Conducting Safety Perception Survey
Displaying BBS banners and posters
Identification of BBS observers for each location and their training by the lead trainers
Survey of present status of unsafe behaviours by lead trainers.
Developing IT enabled BBS observation checklist
Including BBS elements in contract/ tenders terms and conditions
BBS Audit (Internal & External)
Review of BBS implementation and action for improvement by Corporate Task Force
Celebration of BBS implementation and rewarding the best observers based on BBS Best Performance.

9.8 ORGANIZATIONAL STRUCTURE FOR BBS IMPLEMENTATION

The BBS programmes at sites shall include systemic changes, identification of weak areas with respect to physical safety, process safety, electrical safety, etc., and improvement measures to be put in place as per site specific recommendations based on gap assessment analysis report. A successful BBS implementation recommends the following structure (Kaila, 2013).

- Formation of Steering Committee on BBS.
- Formation of Functional Committees on BBS like Committee on Road Safety, Committee on Working at Height, Committee on Electrical Safety, Committee on Process Safety, Safety During Hot Works.

The Functional Committee shall meet at an interval say once in a month and lay down site specific standards, study best practices and implement them after duly approved by the plant head/Officer In-Charge.

- Steering Committee to meet once in two months.
- Roles & Responsibilities of the Steering Committee & Functional Committee shall be laid and discussed.
- Process specific BBS checklist shall be prepared during 5 days roll-out to capture safety data, etc.

9.9 CONCLUSION AND RECOMMENDATIONS

Initially, the concept of safety was like policing and fixing an employee for an accident. It was basically killing the criminal approach not killing the crime. Now we are focusing on unsafe behaviours to be identified and controlled by every trained BBS observer. According to a safety officer, "BBS observer observed some unsafe behaviour which I could not do on routine rounds".

BBS is an organizational intervention to seek best safety results. Unsafe or at-risk behaviours at the workplace mean that the foundation of an accident has been laid. We acknowledge it only when it happens. Unsafe behaviour is a by-product of the work culture. We should never fix the person for his unsafe behaviour, rather continue improving safety culture. A trained BBS observer on an average contributes 2% of risk reduction which precisely means that he/she can save people from accidents at the workplace. People are killed at workplace because we don't alert or save them from their unsafe behaviours. A second of timely correction of unsafe behaviour will save life of your colleagues. 'Feeling for safety of fellow employees' is the foremost in taking next step in safety management (Kaila, 2012, 2013b).

Involving people in observation of live safe / unsafe behaviours is a new way of looking at safety. Take a challenge to correct unsafe behaviours of people around and practice it daily. BBS forms a strong bridge to connect all safety procedures to reach final destination, i.e., zero unsafe behaviour. The success of BBS comes from a 'very strong drive by the management team' to emphasize upon the implementation of BBS across an organization in a routine fashion. The safety officer/manager must convince the top management about safety systems implementation. At the same time, the management should be ready to get convinced about the same but it also depends upon the technical competence and daring of the safety manager to emphasize the same.

It is true that India, as predicted by many, is to become the world's third largest economy by adopting the best work systems such as behaviour based safety for developing no-injury culture. Organizations that fail to take a scientific approach to safety's human-behaviour element are gambling with their futures and are ultimately only safe by accidents. Organizations that adopt behaviour based safety approach move toward zero accident by focusing on 'zero unsafe behaviour' (which is the root cause of any incident) by involving all employees.

REFERENCES

Broadbent David (2014). Email communication.

Fred R. (2014). Email communication.

Inaugural Ceremony of BBS implementation on 11 Dec 2013 at GAIL Delhi.

Kaila H.L. (2012). Behaviour-based safety in organizations, Prasad Psychological Corporation. New Delhi.

Kaila H.L. (2013). Industrial Deaths: Points of concern for insurance companies. The Journal of Insurance Institute of India, Vol. 1 (3), Oct-Dec 2013, 40-48.

Kaila H.L. (2013a). Re-thinking human safety – Is industry really Poised for zero accident? – A review. *Prabandhan Indian Journal of Management*, Vol. 6 (10), 5-12.

Kaila H.L. (2013b). Is Industry Really Poised for Zero Accidents? World Focus, Vol. 13 No. 2 issue, International Practice Specialty's technical publication.

Kaila HL (2013b). Review of BBS Implementation in Indian Organizations. World Focus, Vol. 13 (1), 1, 10-16. International Practice Specialty's technical publication.

10

Re-Contextualizing Safety Culture through Behaviour-Based Safety in Industry

Indian corporates have now recognized that the unsafe behaviours of employees greatly hit the company's business and its balance sheet, and it is hoped that people at workplace would put their efforts to prevent the pain, suffering and financial losses brought about by accidents and related injuries at work (Health and Safety Executive, 2005). While inaugurating a safety seminar, a plant manager of an oil industry stated, "our productivity is going to hamper if we don't efficiently implement behaviour safety practices. We should be recognized as a safe organization". Another safety manager said, "sustainability being a core value of organizations today, behaviour safety approach provides us the best sustainable safety for our fellow employees and workmen". Another manager expressed, "safety mostly suffers as an attachment to the contractor, hence the financial responsibility should be of the contractor but safety responsibility must be held with the company" (Chakraborty, 2016).

A group of 27 safety officers and 20 plant heads in a workshop collectively listed a set of safety compromises they allow and violations that the employees practise on daily basis as below:

- Production targets contradict safety.
- Work permit rules are violated.
- Global/horizontal corrections are not done.
- Top management is more concerned about sale, not safety.
- Pressure from seniors that load had to go for marketing.
- "No safety – no operations" is a clear instruction from the top management but it is not practiced, as the sync between top, senior, middle, junior employees is lacking.
- PPE are given by the contractors but the quality of PPE is compromised often.

- Officers and contract workmen are checked differently by the security personnel at the gate. Seniors are not checked but workmen are thoroughly scrutinized.
- Speed limit of vehicles by company officers is not complied.
- Safety briefing before issue of work permit is overlooked/ not practised.
- Production capacity as per OISD guidelines is not followed. Production often exceeds the plant capacity.
- Mock drills and night inspections are compromised for production targets.
- "Unsafe is ok" is the unconsciously driven mindset in Indian industry that needs a change through BBS approach.

In a concluding session of BBS lead trainers programme on 4th June 2015 (Personal Communication, 2015), the CEO of an Indian copper company emphasized that sometimes it happens that when an observer keeps reporting unsafe behaviours in his department, the HOD may indirectly ask him to suppress it, or it becomes a conflict between the both which is a serious issue, as by not highlighting an unsafe behaviour, one is risking the life of other colleagues. Hence, it is significant that the observers and lead trainers don't hide the at-risk behaviours during discussion in monthly BBS meetings for their rectifications. He empathically mentioned, "let us face this issue, and save lives".

Indian industry is now witnessing a positive transformation in safety management. The behavioural trends across 9 sites in India during the year 2015 (Table 10.1) indicate the varying degree of safe and at-risk behaviours that exist in industrial settings. The correction of at-risk behaviour varied from 30% to 100% and the safe behaviours after correction ranged between 86% to 100% that shows a healthy effect of BBS intervention across sites (Kaila, 2014a).

This chapter has covered the following aspects about the full implementation of BBS towards total injury prevention reviewing data of Indian companies implementing BBS at their sites (Kaila, 2013, 2015).

1. Concerns of BBS Steering Committee.
2. Reduction in Injuries 6-Months Post-BBS Implementation.

Table 10.1: Behavioural Trends across 8 Sites

Location	Number of observations	Safe behaviour	At-risk behaviour	Corrected behaviour	Correction of At-risk behaviour (%)	Safe behaviour (%)
1	14	10	4	4	100%	100%
2	837	641	196	167	85%	97%
3	5066	4336	730	482	66%	95%
4	727	590	137	128	93%	99%
5	154	118	36	18	50%	88%
6	392	282	110	94	85%	96%
7	661	506	155	119	77%	95%
8	161	128	33	10	30%	86%

3. The BBS queries and concerns of CMD.
4. BBS review after 1-year of implementation.
5. Conclusion and implications for practice.

10.1 CONCERNS OF BBS STEERING COMMITTEE

A construction site in-charge speaks out, "we compromise safety due to financial and time pressures from the senior management, as a result of which, I nearly missed a serious accident". Health and safety has become a major concern of industrial organizations all over the world. The key elements shared at the large Gas Company's BBS Project Corporate Meeting on March 2015 were: the BBS project is a Change Management from Culture 1 to Culture 2.

The major concerns of the BBS Steering Committee included as below:

1. Whether HoDs are aware of the existing at-risk behaviours?
2. Is BBS really becoming a line function?
3. Are motivational awards for observers/lead trainers in place?
4. Is BBS a part of the monthly operational review as 20% behavioural risk exists for operations?
5. In 6 months, 38% injuries decreased, safe behaviours went up from 70 to 80%, number of observers increased but BBS training is not fully completed for all employees/ contract workmen.
6. What is the action plan for achieving zero at-risk behaviour for HoDs?
7. BBS banners are still not visible in the plant areas.
8. Quarterly external review is needed for BBS follow-up by the expert.
9. BBS sustainability is simple if we focus on zero compromise/ tolerance for at-risk behaviours. Regular refresher training for existing BBS observers is needed for keeping the momentum on.

Some of the remarks made by the senior managements across Indian work sites are: safety culture is being improved in many aspects as a result of BBS implementation. BBS has made a positive impact on the day-to-day working by way of enhanced safety consciousness and concern among majority of employees including contract employees towards safe behaviours.

In overall, there is a significant change in safe behaviours and involvement of employees and contractors. The employees are committed for building safer organization marching towards zero incidents and no harm culture for which BBS must sustain in the company. Continuity of BBS approach is the success mantra for total safety culture (Kaila, 2013a).

With BBS awareness, you are trying to enhance your perception about the presence of something that happens just before an accident occurs, which is an unsafe behaviour, and BBS training helps an organization to remove this, so that the work environment is safe, everybody is safe, and the production, business, etc., go smoothly. Moreover, when you close your eyes to an unsafe behaviour, then you open your eyes to a fatal accident. BBS is an off-shoot of behavioural science research (Kaila, 2014b).

10.2 REDUCTION IN INJURIES: 6 MONTHS POST-BBS IMPLEMENTATION

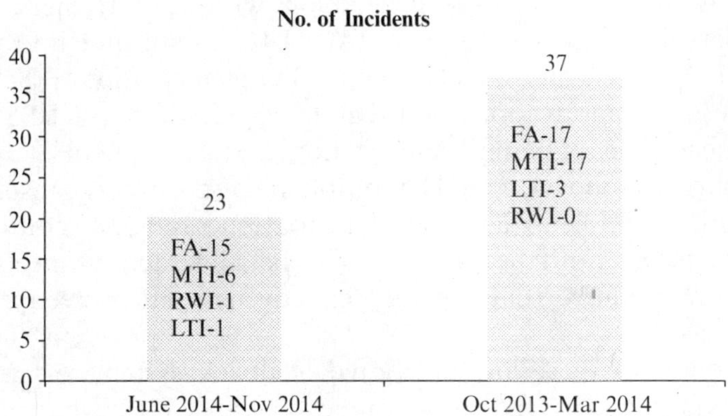

There was a 37% reduction in injuries in 6 months post-BBS implementation in an aluminum unit. The company is benchmarking across departments/locations to targeting zero at-risk behaviour.

10.3 THE BBS QUERIES & CONCERNS OF CMD (CHAIRMAN & MANAGING DIRECTOR)

Query: 1. What is the critical takeaway from BBS journey?
Reply: The critical takeaways from BBS journey are: undoubtedly the heightened involvement of people and clarity about the criticality of unsafe behaviours at site and its spot-correction. However, it needs to be further drilled down amongst contract workmen that need regular emphasis and accountability by the HoDs which is evidently less.

Another outcome of BBS is the declining trend in unsafe behaviours. Critically speaking, the major take-away would be to compare and comprehend the safety incidents record before and after the launch of BBS. This is very significant for reaching towards zero injury and zero at-risk behaviour at sites. However, this would take a couple of years.

Query: 2. Is there any internal assessment being done on the effectiveness of BBS programme?
Reply: The internal assessment is done on the effectiveness of BBS programme through monthly meetings of the BBS steering committees and the functional committees. What weakens the effectiveness of BBS programme is the passive leadership of HoDs and their weekly interaction with lead trainers. The leadership for safe behaviours from EICs is still not up to the mark. What further weakens the BBS implementation is the lack of sufficient motivation, reward and recognition for the BBS lead trainers. The ongoing identification and assessment of gaps through six follow-ups and hand-holding by the expert consultant ensures complete BBS implementation and its continual effectiveness. Both the qualitative (the field interactions/meetings) and quantitative (data driven) effectiveness index need equal emphasis.

Query: 3. Is there any individual feedback system for people undergoing this BBS programme, in case it is, what is the feedback?
Reply: The feedback is the spot correction and counselling on the unsafe behaviour observed and safe behaviours appreciated by the trained BBS observers. However, all trained observers are not active, hence this individual feedback system is weak as

because there is a lack of weekly interactions between lead trainers and observers to regulate this feedback process. What would really strengthen the BBS performance in the organization is the monthly BBS observations/feedbacks essentially being linked to the individual KRA/KRO of employees. The corporate monitoring through monthly reporting from the participating sites in the BBS project serves as a strong tool and the real time feedback mechanism to sustain its implementation. Finally, the continual leadership of top management over the years would assure and sustain full implementation of BBS and the practice of safe behaviours by all sites (Kaila, 2015).

10.4 BBS REVIEW AFTER 1-YEAR OF IMPLEMENTATION

BBS is a whistleblower on the behavioural risks which are apparently business risks like violating standard procedures and so on. BBS means when you are walking through your workplace and observe that somebody is working with such a risk, you stop for a couple of minutes and alert him for spot correction as a big brother not as boss, and this correction procedure is developed in the organization as a regular way of work life involving everyone from top to bottom, and this entire BBS activity is measured in a scientific way each month (Kaila, 2015).

BBS review after 1 year of implementation in an iron ore business organization reveals as below:

	2013-14	**2014-15**		**2015-16 till August**	
	Before BBS	**After BBS**	%	**After BBS**	%
Near-miss	44	16	64	22	50
Incident	50	22	56	13	74
FAI	28	17	39	13	54
MTI	2	1	50	0	100
LTI	3	1	67	0	100
Fatalities	1	0	100	0	100
Total	**128**	**57**	**55**	**48**	**63**

Decrease in Total Number (At-risk acts)	71	80
Increase in safe acts after BBS %	55	63

- There is a reduction in LTI, MTI, and first aid injury. At-risk acts have gone down by 80% and safe behaviours increase by 63%.
- 'Safety is not my concern' has reduced to almost 50%. At-risk behaviours are increasingly being identified. Regular meetings of the BBS steering team are happening. BBS data are being entered in SAP system.
- Reporting of near-miss is encouraged. But when near-miss was reported and a HoD was punished, then people stopped reporting near-misses. Also, as the data entry is by name of observee/defaulter, the observations have gone down from 1200 to 400, i.e., 3:1. Only 40% mentors/observers are active, hence there is a need to refresh and reflect upon the basics of BBS to practice for passive members.
- The company is providing resources if the contractor fails. A weekly observation round by each HoD is happening. Workmen ask for PPE before starting the work. Staff and workmen have started alerting seniors for their unsafe behaviours.
- The BBS drive is there but worker-to-worker (WTW) corrections are not seen. The mentors have to see that the observers training is effective, and it becomes learning for implementation, sharing by observers takes place on weekly basis.
- A safety officer expressed that we allow some minor unsafe behaviour which we should not do. This is in fact atypical of safety officers in most organizations.
- BBS is included in PAP (principles of accident prevention) and PEP (preliminary elaboration of procedures) talks.

10.5 CONCLUSIONS AND IMPLICATIONS OF PRACTICE

Generally, people in India are more than casual in safety precautions while crossing roads, travelling in overcrowded public vehicles, not following road safety rules, lack of safety education in schools, etc. However, Safety Management Leadership across India has significantly increased due to BBS revolution. There is much pressure on senior managements for implementing safety in plants. According to a general manager, "So far, BBS was happening as a corporate mandate to find out who is at fault and doing unsafe and take appropriate action, now we have refined BBS approach as a corrective and proactive way. Today as per Board of Directors, safety is on priority list."

This research points out that the safety officers and site engineers allow several safety compromises, as a result of which the violations that the employees practise on daily basis impact the company's business. BBS postulates that it's not only keeping our plant accident-free, but also society free of accidents as well. We should not be a part of accident statistics rather be a part of 100% safe behaviour. The BBS system requires corporatization so that individual plants get integrated and continuous contribution gets appreciated. The ratio of how much a manager behaves unsafely in personal life as compared to work life, on an average, is almost 30:70. Hence, there is a reason to emphasize upon self-corrections. One of the plant managers said, "Earlier I did not give time to safety officer, now I am trained in BBS, we both would implement BBS". An active leadership matters a lot in creating an accident-free plant (Kaila, 2014).

There is a direct correlation between percentage of employees/ workmen's involvement and percentage of safety culture. If you involve 50% of employees, you get 50% of safety culture. If you wish to have 100% safety culture, you need to involve 100% employees/contract workmen in observation and spot-correction of unsafe behaviours. As OHSAS: 18001 stated that all organizations must adopt a procedure on human behaviour aspects of safety by involving all employees so that they identify and control the risk at workplaces. The accuracy of observations are necessary in corrections of at-risk behaviours, hence employees must have knowledge about their jobs, hazards, SOPs, processes involved, etc. BBS means to fill the missing link between an unsafe behaviour towards safe behaviour of your colleague working with you. A general manager said, "if they (employees) are safe, we are safe, safety is not only mandatory, it's a habit/ behavioural pattern, safety trauma can be reduced the BBS way'. Our final goal is to make BBS everyone's life value, hence it is imperative to activate this dormant value of life saving in every human being in order to create safe societies and safe work life for all.

BBS is a significant contributing tool for corporate safety management performance. An implementation of behavioural safety is meant to a reconstruction of safety value across the corporate towards human life safety. BBS is a permanent solution for developing an assured safety culture involving all employees. It's a change/difference management from cuture-1 to culture-2 which involves changing the present antecedents as depicted in the ABC Model below:

Antecedents (A)	Behaviour (B)	Consequences (C)
Antecedents are the safety systems, policies, procedures that trigger / activate behaviours in the making of an existing culture (C1). BBS builds new safety culture involving all (C2).	Behaviour can be safe or unsafe, and is influenced both by the antecedents & consequences.	Negative consequences are injuries, fatalities, near-misses. Recognizing and rewarding safe behaviours are positive consequences which must be soon, sure, sincere and significant. Individual perception of consequences decides safe or unsafe behaviour.

The key learnings of BBS implementation across industries emerged as follows: every trained employee is a BBS mentor, identification of unsafe acts / conditions becomes a routine activity, reduction of incidents due to timely intervention by observer, safe behaviour a part of DNA, and BBS is the framework for sustaining OH&S Management Systems towards creating safer workplace. BBS is truly a safety-in-action procedure, as a result of which, a monthly BBS scoreboard is displayed in the plant as feedback to all employees. Finally, safety leadership through BBS system can totally transform performance of a business, taking it to newer heights. The management team must show enthusiasm in behavioural observations on continued cycle in sustaining the program (Kaila, 2014c).

To conclude, if we hope to put an end to the human, financial, production and business losses due to accidents at work areas, then we need to essentially empower our human resources to get rid of their at-risk behaviours through BBS implementation and sustain it by involving all of them. To re-emphasize and caution all concerned, it is easily predictable that the workplaces that have not implemented behaviour safety monitoring must expect minor or major incidents repeating time and again, as hundreds of unsafe behaviours go on un-rectified on daily basis (Kaila, 2016).

REFERENCES

Chakraborty J. (2016). Behaviour based safety and you. Management Guru: Journal of Management Research, Vol. 4 (1), 136-141.

GAIL India (2014). Launching of centralized portal on behaviour based safety. December 12, 2014. New Delhi.

Health and Safety Executive (2013). Why leadership is important. Retrieved on June 30, 2013. http://www.hse.gov.uk/leadership/whyleadership.htm

Health and Safety Executive (2005). Good health and safety is good business. Retrieved on June 9, 2015. http://www.hse.gov.uk/costs/costs_overview/costs_overview.asp

Kaila HL. (2010), Behaviour-based safety programs improve worker safety in India. Ergonomics in Design. Vol. 18 (4), Fall, 17-22. CA, USA.

Kaila, H.L. (2013). Review of BBS implementation in Indian organizations. World Focus, 13 (1), 10-16.

Kaila, H.L. (2013a). Is industry really poised for zero accidents? World Focus, 13 (2), 14-19.

Kaila, H.L. (2014). Emerging issues and outcomes of behaviour based safety implementation. Industrial Safety Review, June Issue, 80-86.

Kaila, H.L. (2014a). A case of behaviour based safety implementation at a multinational organization. Journal of Organization and Human Behaviour, Vol. 3, issue 2 & 3, 1-8.

Kaila, H.L. (2014b). Implementing behaviour based safety: Issues and challenges. International Journal of Management Research, vol. 5(1&2), 39-48.

Kaila, H.L. (2014c). Implementing new BBS programs in organizations. World Focus, 13 (3), 1, 8-11.

Kaila, H.L. (2015). Field learnings from Behaviour Based Safety implementation: Reviews in Indian Organizations. Indian Journal of Training and Development, Vol. 45 (4), 53-61.

Kaila, H.L. (2016). Behaviour based safety management across industries: India Survey. Industrial Safety Chronicle, April-June Issue, Vol. XLVII (1), 95-101.

Personal Communication (2015). A meeting with GM (operations) of a multinational aluminum company in south India.

11

BBS Implementation at an Indian Multinational Organization

The success of BBS is truly considered only when the safety department has handed it over to the line management. Safety department should initially involve in coordinating its activities but gradually pass on behavioural safety ownership to the line function. Behavioural safety is being applied successfully worldwide instead of command-and-control approach to occupational safety (Geller, 2004).

BBS is an art of communication and correction of unsafe behaviour at work is the responsibility of one and all to develop zero harm and safe environment. If an unsafe behaviour exists at one place, be aware that the same unsafe behaviour is also prevailing at other locations as it is a reflection of safety culture which requires global or horizontal prevention. For example, if a work-permit or SOP violation is noticed in one unit, it needs to be identified and corrected at other units of the company as well in order to develop safe culture (Kaila, 2014).

Safety first is only a slogan in Indian companies; it does not appear in practice. In realty, its production first, and safety is put to back stage for cost reasons. BBS is much more than enforcement, it is deeper. BBS is yoga for safety. We shall ask each contractor to appoint safety mentor and the company BBS mentor would train them. These are some comments made by participants of this case study.

A major objective of this case is to gain knowledge for organizations considering to implementing BBS towards achieving zero unsafe behaviour targeting zero injury at workplaces.

The BBS programme was implemented in four locations in India at Sesa Sterlite, a multinational company (employing 9000 employees/workmen in its metals and mining business units) that were trained as BBS mentors and observers over a period of one year. This programme included the concept and process of BBS, plant visits for observation and correction of at-risk behaviours, developing road map for implementation of BBS, linking BBS with safety systems such as hazard identification, accident investigation and safety audits, developing training module for imparting training to all employees, formation and functions of BBS steering team. The participants included managers, heads of departments, contractors, and safety officers from across entire plant.

11.1 QUESTIONS MANAGERS PONDERED OVER BEFORE BBS IMPLEMENTATION

Safety is an integral part of all our functions, but even after having all the safety systems in place we are unable to achieve injury-free culture (Kaila, 2013). The managers in the present study viewed the success of BBS with critical perceptions. A set of 20 important questions were asked and clarified by the participants before BBS implementation as below.

1. How do you define BBS approach and what is the benefit of BBS implementation?
2. Why do you want to teach behavioral safety to a person working for 20 years?
3. How to activate or motivate the passive observers? How do we ensure that the observers don't fabricate the observation data?
4. BBS helps the company in production (without accidents), should it not reflect in employee salary also.
5. What is the guarantee that after BBS implementation, accidents would not happen?
6. Are we not creating fear in the minds of workmen by educating consequences of accidents, and that if you don't behave safe, you would get injured or killed?
7. Do we have any guidelines how do we go about implementing BBS?
8. We have many safety programmes, fatalities are not decreasing; how would BBS help?

9. Does it mean that earlier safety programmes have failed, that is why we are doing BBS?
10. Would BBS implementation increase our work, i.e., training more observers, etc?
11. Is BBS a magic wand that by introducing it, we are going to achieve a safe environment?
12. Do we need to improve other safety systems such as contractor management, HIRA, safety audits along with BBS? How shall we sustain BBS?
13. Is there any benchmarking for BBS progress or behavioural trends?
14. How do you deal with difficult/tough/high risk observees?
15. What are the two challenges of BBS implementation?
16. Do we need additional time or extra efforts to implement BBS?
17. How to achieve rise in safe behaviours and fall in unsafe behaviours?
18. What is the safety status before and after BBS implementation?
19. Whether procedural deviations are linked to supervisors and managers behaviours?
20. How to control the most critical incidents with BBS approach?

Each of the above questions /doubts was addressed in details to inspire the mentors to implement BBS. If these apprehensions are not clarified, it creates problems in implementing.

11.2 CRITICAL BEHAVIORS RESPONSIBLE FOR ACCIDENTS

The existing data of the company indicated that besides other factors, the major reason for the incident occurrences happened due to not following safety procedures (32.88% and 57.89%), i.e., LOTO, SOP, work permit system (WPS).

Critical Behaviours	Year 2013-14		Year 2014-15	
	Occurrence	%	Occurrence	%
PPE	11	7.53	1	5.26
Housekeeping	10	6.85	1	5.26
Using Tools & Equipment	20	13.70	1	5.26
Body Positioning	7	4.79	2	10.53
Material Handling	6	4.11	1	5.26
Communication	1	0.68	0	0.00
Following Procedures	48	32.88	11	57.89
Visual Focusing	34	23.29	2	10.53
Use of Mobile at Work	1	0.68	0	0.00
Others	8	5.48	0	0.00
Total	146	100.00	19	100.00

In addition to the above, the experienced managers from different sites identified common accident causes that have rooting to behaviours were as follows:

- Taking shortcuts.
- Being over-confident.
- Starting a task with incomplete instructions.
- Poor housekeeping.
- Ignoring safety procedures mental distractions from work.
- Failure to pre-plan the work.

11.3 BBS ROADMAP

The data of regular plant observation visits by BBS mentors and observers indicated that on an average, the safe behaviours observed were 65%, at-risk behaviours were 35%, and correction of at-risk behaviours were 70%. It was found that after corrections, the safe behaviours increased from 65% to 90% which indicated that the mentors and observers not only learned the BBS concepts; they also practically applied and tested the efficacy of BBS implementation.

The trained mentors and observers prepared a roadmap and action plan along with a responsible person and target date to achieve each action as below in order to implement the BBS approach for all employees and workmen in the organization.

Project Title for BBS Implementation: Zero Unsafe Behaviour & Zero Harm

Sl. No	Actions	Responsibility	Target Date
1	Circular by COO with a message for BBS project implementation	COO office	16 Jun 14
	Forming BBS steering teams for every plant - BBS convener for plant : To be decided by plant head/steering team - Steering team members : Plant head, depts. HoD, safety head	Unit head	14 Jun 14
	Ceremonial inauguration of BBS programme for all units with display of inaugural banners	HR(for function) /Safety(for banners)	18 Jun 14
	Announcing BBS inaugural ceremony on company portal as corporate initiative	IT department-PID	18 Jun 14
	First meeting of unit level steering team at respective units	BBS Convener - Plant	19 Jun 14
	Developing IT enabled BBS observation checklist & creation of group-mentors of BBS (M-BBS)	IT department-PID	14 Jun 14
2	Identifying BBS observers from staff, workmen & contract (10% of workforce) from each areas shift-wise & conducting practical training for the same.	HR/M-BBS	15 Jul 14
	Conducting BBS awareness training for all employees and contract workmen	HR/M-BBS	31 Aug 14

	Displaying BBS banners and posters across areas.	Safety Dept.	31 Jul 14
3	Ensuring daily observation rounds by trained observers & entering daily data in IT enabled checklist	M-BBS & Observers	Daily
	Ensuring strong leadership drive (mentoring, monitoring & motivating) by all HoDs for BBS project activities.	Plant Head/ HoD	Monthly review meeting
4	Compliance with proceedings of steering team (ST) meetings	Steering team	Monthly review meeting
	Setting guidelines for linking BBS with PMS/KRA and HSE system;	HR/steering team	30 Sep 14
	Conducting quarterly interplant BBS Audit by M-BBS	Steering Team	18 Sep 14
	Conducting half yearly external BBS audit by BBS expert.	BBS convener	18 Dec 14
5	Monthly review meeting of BBS steering team with observers	Steering teams	Monthly review meeting
	Display of behavioural trends across plant at AO war room, depts. notice board	Safety depts.	Weekly
	Presentation of BBS trend across various units	Safety Depts.	Monthly
	Comparing safety statistics every six months before /after launch of BBS	Safety Depts.	18 Dec 14
	Celebrating BBS annual day.	HR/Safety Depts.	18 Jun 15
6	Fixing BBS training budget-raising note for approval	Steering team	19 Jun 14
	Issuing BBS sticker for observers' identity to all trained observers;	M-BBS	15 Jul 14

	Including BBS Rules in vendor's terms & conditions (appointment of BBS observers)	Steering team/ Purchase depts.	19 Jun 14
	Rewarding monthly best observers across all categories based on maximum number of observations as per approved scheme by steering team.	Steering team/ HR	Monthly

Note: Time frame for full implementation of BBS – one year and sustaining further.

The plant head, safety head as well as all HoDs discussed the above action plan in greater detail and then ensured to implement in their respective workplaces. The plant head stated that I failed to address the behavioral aspects of safety earlier but now I will do it as per roadmap prepared.

The trained observers from another business unit of the plant (where BBS is implemented) expressed that the company should not only look at the number of observation cards filled-in, we shall go on making corrections of unsafe behaviours, but may not fill cards. For example, during the shutdown of plant, the operators made hundreds of corrections of unsafe behaviours but did not fill-in cards due to lack of time. At the same time, 50% of observers were found to be passive as they made only a few observations in the previous months. There is need to train more observers in order to create an improved safety culture. However, the HoDs emphasized the importance of filling-in observation cards for generating behavioral trends.

11.4 LINKING BBS WITH HAZARD IDENTIFICATION

It is observed that the safety systems, documents, procedures are in place but lack in reflecting safe behaviours. Hence, the need is felt to relate an important safety system like hazard identification with BBS. In this pursuit, the eighteen types of hazards were identified as below and for each hazard, a set of behavioral corrections were also identified which should be ensured by the BBS observers and the BBS mentors must discuss them during

BBS monthly steering team meeting. It is important to identify and control environmental hazards in order to create zero harm culture.

Hazard identification checklist to be ensured by observers & monthly audit by mentors of BBS in each unit

S. No.	Type of Hazard	Behavioural Correction
1	Entanglement	- Following SOP - Proper housekeeping - Avoiding shortcuts - Proper guarding/barricading of moving parts of equipment - No loose clothing, jewellery - Alertness - Avoiding intoxication such as alcohol/ drugs - Proper rest between work hours in shifts
2	Gas Poisoning	- Authorised entry - Following SOP- 2 persons with 2 CO monitors - Following work permit - Ensure healthiness & timely calibration of detection systems - Use of PPE's (gas masks) - Prompt correction of leakages post alarms - Display caution & awareness boards - Not to work with empty stomach
3	Electrical Shock	- Use of certified tools & tackles - Ensure proper shutdown & LOTO compliance - Provide proper earthing - Avoid wet surfaces - Avoid loose cable joints - Caution boards at required places - Do not bypass safety interlocks - Authorisation details on panel rooms - Proper nomenclature of feeders

		- Use of RCCB - Not exceeding no. of joints than specified in SOP
4	Dust	- Dust suppression using water sprays, humidifier, fog gun, etc., - Use of dust masks & PPE - Proper maintenance of de-dusting systems - Enforcing limitation to speed limit of vehicles in plant
5	Working at height	- Compulsory pep talks before start of work - Use of certified ladders, scaffolds, platforms - Use of safety harness, safety net - Maintaining 3-point contact while climbing & getting down - Supervision - Use of mechanized elevated platform/ certified jhoola for the work - Check for overhead live cables at the height of work - Ensuring no-violation of work permit
6	Exposure to high temperature	- Using gloves, eyeglasses, boiler suits, leg guards, face shield, safety shoes, etc., - Use of aprons, overcoat - Body positioning & alertness on the job - No use of gum boots - Adherence to signboards displayed in the area - Work permit compliance - Maintain first-aid box in the vicinity
7	Fall of material/object from height	- Use of safety helmet by the people - use of certified lifting tools & tackles like slings, Chain pulley blocks, D-shackles, etc., - Barricading of area with tape

		- Working at 2 levels (one above the other) not to be allowed, may be allowed with precautions like blanking. - Pre-lifting plans - Proper housekeeping - Check for toe guards & hand rails - Materials not to be thrown - Use of safety net wherever required - Do not leave any unsecured objects on roof/work platforms/walkways -Proper arrangement to carry tools
8	Vehicle movement	- Identify the pathways for transit - one way/2 way - Ensuring adherence to speed limits - Ensure no overtaking/overloading of vehicles - Ensure the driver does not climb the vehicle while loading/unloading - Ensure proper parking methods - Should not cook/rest below the vehicle - Spotter & loaders should wear fluorescent jackets - Random alcohol test - Ensure vehicle is not moving with lifted body - No mobile usage while driving - Looking through the rear view mirror & co-ordination with the spotter - Ensuring queue discipline - Ensure reverse horn - Check for awareness of motor vehicle act by the drivers (PUC, Valid DL, etc.)
9	Noise	- Use of ear plugs/mufflers - Periodic measurement of noise levels - Providing acoustic/silencer at generating source - Providing signages

10	Poor illumination	- Ensuring periodic measurement of lux levels - Ensure adequate illumination before start of work - Improve illumination with additional lights - Keep emergency lights handy
11	Explosion	- Following SOP & work permit - Ensure purging of gas line before work on gas lines - Ensure periodic testing of pressure vessels & gas lines - Keep manholes/relief valves/ inspection covers open & take shutdown while working on gas lines - Ensure use of flashback arrestors on both sides, torch & cylinder side - Dissemination of knowledge of assembly points - Proper storage of flammable substances - No smoking in gas-prone areas - Awareness of material safety data sheet - Use of Flameproof light fittings/ Cables in LPG yard/battery charging rooms/petroleum storage - No storage of gas cylinders in direct sunlight
12	Corroded structures	- Ensure periodic check of structures for stability & corrective action with painting etc. - Prevent overloading of structures - Ensure spillages are cleaned
13	Drowning	- Barricade the areas with potential for drowning - Use of lifebuoys/trained employees to rescue drowned persons - Ensure SOP is followed (at least 2 persons while going to waterbodies)

		- Use of proper pathways - Avoid shortcuts while walking
14	Trapping under coal	- Make proper benches while stacking coal & rectify all undercuts - Correct choice of material handling equipment - Periodic refresher Training of persons using material handling equipment - Barricading of heaps not in use - Ensure proper Illumination & ventilation - Proper access & pathways for heavy equipment movement
15	Reptile bite	- Good housekeeping - Proper illumination - Avoid handling of reptiles, call animal rescue expert - Creating awareness about hazards due to reptiles
16	Job by Untrained Person	- Ensure safety & on the job training under close supervision - Ensure no job on visitor gate-pass
17	Poor Housekeeping	- Ensure 5S is in place
18	Ergonomics	- Educate & create awareness among people

11.5 CERTAIN ISSUES IN BBS IMPLEMENTATION

During BBS implementation, some problems were often cited at workplaces as follows:

- My manager is not giving time for observation, he is overloading me with work.
- My manager is first concerned with production, then safety;
- During emergency situations, we don't follow SOP.
- Managers are focusing on plant production, not concerned about unsafe conditions.
- People often overlook unsafe behaviours thinking that it is small, not critical.

All implementations of BBS would have some ups and downs in its progress (Kaila, 2014). Sometimes employees are not clear about the subtle difference between safe and unsafe behaviours. In a focused-group interview, a team of observers and implementers expressed that we take accidents in a punishment route and warning which BBS is not. The main causes identified for the slowness in developing BBS culture are the lack of regular internal reviews, poor leadership by HoDs, lack of motivation for inactive observers to become active, managers not being convinced or confident of BBS approach or its outcomes, line managers feel that safety is the responsibility/function of safety department, importance to production not safety, lack of an open mind towards peer-to-peer correction, and people feel that it is an extra task so the attendance in BBS meeting is very less. The concept of BBS is not drilled down into the hearts of people. An observer said, "The way we have gone about BBS process is too formal, it's a drama like the observer would inform an observee that he would observe him while he was doing his job". The safety officers stated that during the plant shut-down, about 1200 deviations were corrected by the safety department which means the BBS observers were not doing their duties as expected of them. All this caused loss of interest in BBS project implementation. Any organization planning to launch BBS must look into these practical aspects.

It is important to underline that all observers who contribute to risk-reduction must be appreciated and recognized regularly following a distributed reward approach like ICICI payback system, not that one best observer of the month is rewarded and others don't receive any admiration. Observers also need to know how they are performing in BBS implementation. As one of the observers said, "If observations are happening fast, feedback should also be fast by way of interactive meetings between implementers' team and observers".

A plant head in India expressed that we have failed to learn from our accidents. The second major reason is the lack of value for human life. And the third is the leadership focus on production-not-safety and cost considerations of safety (Health and Safety Executive, 2013).

11.6 BEST EXPERIENCES OF BBS APPROACH

Besides implementation problems, the trained observers listed many learning experiences of BBS approach as below:

- It's a correction of unsafe behaviour and value for human life.
- Be an active observer: never neglect, correct unsafe behaviour.
- Unsafe behaviours that have potential for accidents are allowed with knowledge of employees and BBS is an elimination of root causes of accidents.
- It is a development of additional manpower for EHS team, it's a line function.
- It is an art of safe living, and a humanistic approach, safety for self and others.
- It's a fundamental concept that everybody has right to go back home safely.
- It's more than a safety regulation.
- It's the easiest way to adopt for sending workers injury-free back home.
- In order to create safe culture, each trained observer must review one work-permit a day and at least interact with one contract workman a day.

11.7 BBS LEADING INDICATORS

The following 12 indicators were used to mark the monthly progress of BBS in the company.

1. Number of Observers trained (for 6 hours).
2. Number of contractors trained (for 1 hour).
3. Number of SIIL employees trained (~1 hour).
   - Percentage of total workers trained (employees plus contractors).
4. Number of contractors refresher trained.
5. Number of SIIL employees refresher trained.
6. Number of observation cards completed.

- Ratio of observations this month : total workers
- No. of Safe / Unsafe behaviours with respect to critical behaviours
- PPE
- Housekeeping
- Using tools & equipments
- Body positioning
- Material handling
- Communication
- Following procedures
- Visual focusing
- Use of mobile at work
- Others

7. Number of Observer meetings held.
8. Number of implementation team meetings held.
9. Number of active Observers this month.
    - Per cent company employees active as Observers this month
10. Number of barriers identified.
11. Number of barriers eliminated.
12. Per cent of barriers eliminated.

11.8 CONCLUSIONS: OUTCOMES OF BBS IMPLEMENTATION

BBS implementation across companies revealed the following outcomes (Kaila, 2010).

- Behavioural trends on an average across companies is 72% of safe behaviours; at-risk behaviour of 28%% and at-risk behaviours corrected is 2% by each observer.
- The number of safe behaviours has increased from 60% up to 96% within 6 months of launch of BBS.
- The number of observers has also increased every month.
- The number of unsafe conditions and unsafe behaviours has drastically gone down.

- BBS implementation has been recommended to other locations of the company with an introduction of BBS in one location.
- The management commitment for safety has gone up due to involvement.
- Safety has become a real line function due to BBS approach.
- The incident reporting is not there from locations where BBS is implemented.
- There is a significant decline in near-miss and injuries.
- There is an increased reporting of unsafe behaviours.

The above data has been drawn from many companies (where BBS is introduced for the last 5 years) such as Vedanta, L&T, IOCL locations, India Glycols, Paradeep Phosphates Ltd, Bayer CropScience, Aarti industries, Sarda Energy & Minerals Ltd. (Kaila, 2011).

While concluding the Behaviour Based Safety: "Zero Unsafe Behaviour and Zero Harm" programme, the Chief Operating Officer of the company stated that the company has taken an initiative to launch BBS in our plants with immediate effect for full implementation of BBS approach. Let us all join in this mission of making our workplace safe and healthy. BBS Steering Team has been formed to take this BBS implementation further. In this regard all HoDs as well as employees and contractors are requested to join this safety movement for achieving total safety culture. An excited observer remarked that banners have come and we are in the process of displaying. I am giving BBS training in the plant. It is a great experience to motivate people and its very tough job. I am feeling that I am doing very good job by inspiring people about their safety. I feel that being a BBS trainer, I may save some lives. Needless to say that BBS is a positive intervention/ a correction process, and not a fault finding or blaming procedure. It is an application of Behavioural Science, the purpose of which is to achieve zero unsafe behaviours (Kaila, 2013a).

The lead trainers from all India locations in September, 2016 narrated that the BBS implementation at GAIL brought out a host of positives such as:

- The observers /lead trainers are now involved in training in addition to HSE staff.
- The frequency of safety meetings has increased.
- The safety awareness and compliances have gone up.
- People are working more safely (safe behaviours increased from 65 to 85 per cent).
- There is an increased focus on behaviour and a structured approach is being used.
- There is a distribution of safety ownership. Safety culture has evolved.
- Contractor involvement in safety has boosted.
- There is a shift in safety culture from reactive/dependent to independent/ interdependent.

In order to develop sustainability for BBS at site, a set of 12 actions are suggested (Kaila, 2016):

- Cultural-shift measurement survey from reactive, dependent to independent, interdependent level.
- Accuracy and quality of behavioural index.
- Management engagement (frequency of monthly steering meetings).
- Frequency of observations and corrections.
- Continuity of awareness building and strengthening risk-based conversations.
- Awards, recognition and motivation of observers, mentors and departments.
- Comparing injury/accident data before/after BBS launch.
- Linkage of BBS with employee performance management.
- Departmental leadership commitment at HOD level.
- Difference between BBS data and field reality.
- Propaganda/publicity of BBS at site.
- Extensions to road safety, home, personal life corrections, neighborhood industry.

REFERENCES

Geller, E. Scott (2004). Behavior-based safety: a solution to injury prevention: Behavior-based safety empowers employees and addresses the dynamics of injury prevention. Risk & Insurance, 15 (12, 01 Oct) p 66.

Health and Safety Executive (2013). Why leadership is important. Retrieved on June 30, 2013. http://www.hse.gov.uk/leadership/whyleadership.htm

Kaila H.L. (2010), Behavior-based safety programs improve worker safety in India. Ergonomics in Design. Vol. 18 (4), Fall, 17-22. CA, USA.

Kaila H.L. (2011). Organizational cases on behaviour based safety (BBS) in India. International Journal of Human Resource Management, Vol. 22, 10-12, 2135-46, June-July 2011.

Kaila, H.L. (2013). Review of BBS implementation in Indian organizations. World Focus, 13 (1), 10-16.

Kaila, H.L. (2013a). Is industry really poised for zero accidents? World Focus, 13 (2), 14-19.

Kaila, H.L. (2014). Emerging issues and outcomes of behaviour based safety implementation. Industrial Safety Review, June Issue, 80-86.

Kaila, H.L. (2016). Behaviour based safety management across industries: India survey. Industrial Safety Chronicle, April-June Issue, Vol. XLVII (1), 95-101.

12
संगठनों में व्यवहार आधारित संरक्षा

अपने कर्मचारियों को संरक्षित कार्य स्थान प्रदान करने के लिए व्यवहार संरक्षा एक दवा है। यह नई पीढ़ी की संरक्षा संकृति है। संरक्षा के लिए व्यवहार आधारित संरक्षा बीमा है। पहला लाभार्थी संरक्षा विभाग है जिसे संगठन के संयंत्र में सभी और बीबीएस के कार्यान्वयन से बीबीएस प्रशिक्षित पर्यवेक्षक प्राप्त होते है।

प्रत्येक संयंत्र का एक मुख्य संरक्षा उद्देश्य शून्य दुर्घटना को प्राप्त करना होता है। चूँकि संगठन में संरक्षा 100 प्रतिशत से कम नहीं हो सकती है, इसलिए संरक्षा में प्रत्येक संयंत्र के कर्मचारियों की भागीदारी 100% से कम नहीं होनी चाहिए। इसके बारे में यही व्यवहार संरक्षा है।

यदि हम व्यवहार संरक्षा की संकल्पना और प्रक्रिया के बारे में प्रत्येक कामगार/कर्मचारी को प्रशिक्षित करें, तो हम इस उद्देश्य को प्राप्त कर सकते हैं। संगठन में कर्मचारियों के प्रशिक्षण से पूर्व, हमें विभागाध्यक्षों और फ्रंट लाइन मैनेजरों के बीच व्यवहार आधारित संरक्षा की संकल्पना को सुग्राही या प्रोत्साहित करने की आवश्यकता है।

बीबीएस के कार्यान्वयन से पूर्व, संरक्षा पुलिस निरीक्षण की तरह था। बीबीएस के प्रशिक्षण से अधिकारियों के संचार संरक्षा की ओर कामगारों के दृष्टिकोण में सुधार हुआ है। व्यक्ति बीबीएस को अपना रहे हैं। बीबीएस कार्यक्रम के आंतरिक प्रशिक्षकों और किसी संगठन की प्रशिक्षित टीम के सदस्यों ने काफी अच्छा काम किया है क्योंकि वे कार्य की प्रक्रिया को जानते हैं।

बीबीएस की आवश्यकता

अनुसंधान एवं अनुभव इंगित करता है कि:

1. 90% या इससे ज्यादा दुर्घटनाएं असंरक्षित मानवीय कृत्यों और व्यवहारों के कारण होती है।
2. 50% असंरक्षित व्यवहार किसी भी दिए गए समय पर किसी भी संयंत्र में या चिह्नित किए जा सकते हैं।
3. कर्मचारियों में 25-30% की संरक्षा जागरूकता की कमी है जो कि उनके असंरक्षित व्यवहारों से प्रतीत होती है।
4. असंरक्षित व्यवहार किसी भी बाल-बाल कार्य दुर्घटना और दुर्घटनाओं के केंद्र बिंदु में होते हैं। यदि हम असंरक्षित व्यवहारों को नियंत्रित कर लें, तो हमें कभी भी बाल-बाल बचे दुर्घटना का सामना नहीं करना पड़ेगा।

इसलिए, हमें संरक्षा में संरक्षित और असंरक्षित व्यवहारों पर ध्यान देने के प्रयासों की आवश्यकता है। सफलता का बीबीएस राज यह है कि प्रत्येक कर्मचारी के हाथों में संरक्षा नियंत्रण हो, वे स्वयं को जिम्मेदार और स्वस्थ महसूस करें।

ओएचएसएएस का पालन ही बीबीएस है

ओएच एंड एस प्रबंधन प्रणालियों में ओएचएसएएस 18001:2007 दस्तावेज शामिल है : भागीदारी एवं परामर्श धारा 4.4.3.2 कि संगठन खतरों को चिह्नित, जोखिम का निर्धारण करने और नियंत्रणों का निश्चय करने में कर्मचारियों की भागीदारी और उनकी उचित अपेक्षा की प्रक्रिया(प्रक्रियाओं) को सत्यापित, कार्यान्वित और अनुरक्षण करेगा।

ओएचएसएएस 18001:2007 के खंड 4.3.1(जोखिम पहचान की योजना, खतरा निर्धारण और नियंत्रण निश्चित करना) के अनुसार कि संगठन मानवीय व्यवहार, क्षमताओं और अन्य मानवीय कारणों को शामिल करते हुए खतरों को चिह्नित और जोखिम का निर्धारण करने के लिए प्रक्रिया(प्रक्रियाओं) को स्थापित, कायान्वित एवं अनुरक्षण करेगा।

ओएचएसएएस 18001:2007 के खंड 4.3.2(सक्षमता, प्रशिक्षण एवं जागरूकता) के अनुसार कि संगठन उसके नियंत्रणाधीन कार्य कर रहे व्यक्तियों को ओएच एंड एस के वास्तविक और संभाव्य परिणामों, उनकी कार्य गतिविधियों, उनके व्यवहारों और समुन्नत व्यक्तिगत निष्पादन के ओएच एंड एस के लाभों से अवगत कराने के लिए प्रक्रिया(प्रक्रियाओं) को स्थापित, कार्यान्वित एवं अनुरक्षण करेगा।

बीबीएस प्रक्रिया में ओएचएसएएस 18001:2007 को शामिल किया गया है(इसके प्रशिक्षण अनुप्रयोगों में मानवीय कारणों का अनुपालन)। व्यवहार आधारित संरक्षा संगठनों को ओएचएसएएस 18001:2007 के उपर्युक्त खण्डों के कार्यान्वयन में मदद करेगी। बीबीएस पद्धति खतरों को चिह्नित करने, जोखिम का निर्धारण करने और नियंत्रणों का निश्चय करने में कार्मिकों की संबद्धता और भागीदारी को सुनिश्चित करना है और बीबीएस कार्मिकों को उनके असंरक्षित के साथ ही साथ संरक्षित व्यवहारों के बारे में भी अवगत कराना है। बीबीएस मासिक आधार पर कार्मिकों के असंरक्षित और संरक्षित व्यवहारों का प्रतिशत के रूप में हिसाब रखने में भी सहयोग करेगा।

ओएचएसएएस 18001:2007 के उपर्युक्त तीन खंडों के कार्यान्वयन बीबीएस सहायता करेगा।

बीबीएस प्रशिक्षण उद्देश्य

बीबीएस पर यह एक-दिवसीय कार्यशाला बीबीएस संकल्पना, टिप्पणी और फीड बैक प्रक्रिया और किसी संगठन में बीबीएस के कार्यान्वयन के पहलू और अनुभव को बताएगी।

संगठनों में दुर्घटनायुक्त संस्कृति के विकास के लिए संरक्षित व्यवहारों को प्रोत्साहित करने और दुर्घटनाओं में कमी के लिए असंरक्षित व्यवहारों को सुधारने के संबंध में यह कार्यशाला व्यवहार आधारित संरक्षा की प्रक्रिया और संकल्पना के अनुप्रयोग की गहन समझ पर विशेष बल देती है।

कार्यशाला विषयवस्तु

संगठनों में बीबीएस को प्रारंभ करने से पूर्व, कर्मचारियों के प्रशिक्षण पर गंभीरतापूर्वक विचार किया गया। अतः बीबीएस प्रशिक्षण कार्यशाला में निम्नलिखित विषय-वस्तु को शामिल किया जाना चाहिए:

9.30 बजे से 4.30 बजे तक
• व्यावहारिक संरक्षा की संकल्पना। • व्यावहारिक संरक्षा की प्रक्रिया • संरक्षित एवं असंरक्षित व्यवहारों का मापन। • संरक्षित एवं असंरक्षित व्यवहारों का प्रदर्शन एवं दस्तावेजीकरण। • संगठन में सभी ओर व्यावहारिक संरक्षा का कार्यान्वयन। • बीबीएस के लिए कार्य योजना/रोडमैप।

व्यावहारिक संरक्षा प्रशिक्षण क्या करता है ?

बीबीएस एक सर्वोत्तम और अद्यतन संरक्षा पद्धति है। किसी कर्मचारी के पास बार-बार जाना और जब तक वह संरक्षित व्यवहार और स्व-अवलोकन की संकल्पना को नहीं सीख लेता तब तक या ऐच्छिक रूप से अवलोकन करना और संरक्षित निष्पादन के लिए अन्य लोगों के अवलोकन की प्रक्रिया ही बीबीएस है। असंरक्षित व्यवहारों को रोकने और संरक्षित व्यवहारों को प्रोत्साहित करने के क्रम में व्यावहारिक संरक्षा निम्नलिखित आठ पहलुओं को शामिल करती है:

1. पी पी ई
2. हाउसकीपिंग
3. औजारों एवं उपकरणों का प्रयोग

4. बॉडी पॉजिशनिंग/प्राटेक्टिंग
5. सामग्री का रख-रखाव
6. संचार
7. प्रक्रियाओं का पालन
8. दृश्य(vissual) पर ध्यान देना

चेकलिस्ट की सहायता से, बीबीएस के प्रशिक्षित पर्यवेक्षक अपने अवलोकनों को निर्देशित करने के लिए उपर्युक्त आठ क्रिटिकल व्यवहारों के लिए डाटा तैयार करने हैं। प्रत्येक पर्यवेक्षक कम से कम एक सहकर्मी का रोजाना अवलोकन करता है। इस प्रकार संरक्षा एक दैनिक अनुस्मारक के रूप में स्थापित हो जाती है जोकि सारांशतः संरक्षा संस्कृति का निर्माण करके हम लगभग रोजाना असंरक्षित और संरक्षित व्यवहारों में हुई बढोतरी या कमी की जॉच कर सकते है।

बीबीएस इस बात पर जोर देता है कि कर्मचारियों को उनके संरक्षित व्यवहारों के साथ ही साथ असंरक्षित व्यवहारों की जिम्मेदारी लेने की आवश्यकता नहीं है। यदि वे असंरक्षित व्यवहार करते है तो उन्हें दंडित करने के बजाय बार-बार सही करने के लिए कहा जाएगा और जब से संरक्षित व्यवहार करते है तो उन्हें प्रोत्साहित किया जाएगा। असंरक्षित और संरक्षित दोनों ही तरह के व्यवहारों की गणना एवं प्रदर्शन किया जाएगा। बीबीएस असंरक्षित व्यवहारों को प्रभावित करने वाली असंरक्षित परिस्थितियों के बारे में भी चर्चा करता है।

बीबीएस एक डाटा-आधारित निर्णय लेने की प्रक्रिया है। बीबीएस का विकास है कि जो कुछ भी माप लिया गया है उसको पूरा कर लिया जाए और संगठन की संरक्षा में प्रत्येक कर्मचारी विशिष्टता ला सकता है। व्यावहारिक परिवर्तन(अवलोकन एवं सही करना) में विशेषज्ञता के लिए कर्मचारी ही मूल स्रोत है। समस्त कार्य क्षेत्रों और विभागों हेतु ब्रीफिंग सत्रों के साथ बीबीएस प्रारंभ होता है। बीबीएस एक टीम वर्क है, यह पूरी कंपनी में होता है और लोगों द्वारा चलाया जाता है। संरक्षा नियमों को लागू करना, परिवर्तन पर जोर देना, दूसरों के बारे में गपशप करना, बॉस को रिपोर्टिंग करना बीबीएस का उद्देश्य नहीं है। संरक्षित और जोखिम व्यवहारों की पहचान, चोट की संभावना की पहचान, जोखिम का संचार और संरक्षित समाधानों की पहचान करना इसका उद्देश्य है। एक कार्यान्वयन टीम अथवा बीबीएस संचालन समिति इसकी प्रगति पर नजर रखेगी। मूलभूत रूप से बीबीएस संरक्षा के लिए प्रबंधन की ओर से चलाया जाने वाला औजार नहीं है। यह प्रबंधन के सहयोग से कर्मचारियों द्वारा चलाई जाने वाली पद्धति है।

बीबीएस की विशिष्टताएं

ये विशिष्टताएं बीबीएस की भिन्न-भिन्न विशेषताओं को शामिल करती है जो कि बीबीएस में शामिल संकल्पनाओं और पद्ध्तियों को समझने में हमारी सहायता करती है:

ए) बीबीएस संकल्पनाएं

बी) अवलोकन एवं फीडबैक प्रक्रिया

सी) बीबीएस कार्यान्वयन एवं संचालन समिति

बीबीएस संकलपनाएं

1. असंरक्षित एवं संरक्षित व्यवहारों का सक्रिय रूप से ध्यान रखती है जिससे संरक्षा व्यवहार में सुधार होता है।
2. सुनना, प्रशंसा करना, मिलकर समस्या का समाधान ढूंढना और संरक्षा उपलब्धियों पर खुशी मनाने से सावधानीपूर्वक व्यवहारों में सक्रिय रूप से बढ़ोतरी हो सकती है।
3. दृष्टिकोण और व्यवहार संबंध कमजोर है। जो कुछ भी हम पढ़ाते हैं वह आवश्यक रूप से व्यवहार में परिवर्तित नहीं होता है।
4. दष्टिकोण आंतरिक है जो सोच और अनुभूति से संबंधित है, जबकि व्यवहार बाहरी है जिसका अवलोकन और सक्रिय अनुभव किया जा सकता है।
5. बीबीएस संरक्षा के लिए व्यक्ति और समाज को संबोधित करता है।
6. बीबीएस प्रक्रिया मूल्यांकन, घटना विश्लेषण अथवा पर्यावरणीय समाधानों को प्रतिस्थापित अथवा बदलता नहीं है। जब जोखिम व्यवहारों की पहचान कर ली जाती है तब पर्यावरणीय कारणों पर टिप्पणियां की जाती है ताकि ऐसे व्यवहारों को कम अथवा रोका जा सके।
7. संरक्षा निष्पादन में हुए परिवर्तन को समझने के लिए बीबीएस के कार्यान्वयन से पहले और बाद में संरक्षा वातावरण अथवा जागरूकता का माप करके बीबीएस के प्रभावों को देखा जा सकता है।
8. आपके संगठन की आवश्यकताओं के अनुसार बीबीएस को प्रचालित प्रक्रिया की आवश्यकता होती है।
9. बीबीएस ''करें'' अर्थात: परिभाषा, अवलोकन, हस्तक्षेप और परीक्षण का पालन करता है।
10. बीबीएस ने कई देशों एवं उद्योगों में संरक्षित व्यवहार और दुर्घटनाओं की दर में कमी के संबंध में सकारात्मक परिणाम प्रदर्शित किए हैं।

11. बीबीएस संरक्षित व्यवहारों में वृद्धि तथा चोट, बीमारी और संबंधित वित्तीय लागतों में कमी करता है।
12. बीबीएस में अवलोकन और फीडबैक की एक प्रक्रिया, आंकड़ों के समीकरण, विश्लेषण और प्रचार की एक प्रणाली और प्रबंधन सहयोग की प्रोएक्टिव पद्धति शामिल है।
13. बीबीएस सक्रिय रूप से ध्यान रखता है, यह स्वयं और अन्य के लिए संरक्षा हेतु कॉल ऑफ ड्यूटी से भी ज्यादा ध्यान रखता है।
14. बीबीएस प्रत्येक के लिए संरक्षा है।
15. बीबीएस एक सतत् प्रक्रिया है जब तक कि आप कार्यस्थल पर असंरक्षित व्यवहारों को रोकने का निश्चय नहीं कर लेते।
16. संरक्षित व्यवहारों के लिए बीबीएस समकक्ष व्यक्तियों से सीखना है।
17. असंरक्षित व्यवहारों में कमी के लिए बीबीएस निश्चयात्मक प्रगति की एक प्रक्रिया है।
18. बीबीएस संरक्षा विनियमों से कहीं ज्यादा है।
19. बीबीएस कोई चमत्कार नहीं है। यह एक सतत् प्रक्रिया है।
20. बीबीएस कोई दण्ड अथवा अनुशासनिक कार्रवाई अथवा दुर्घटना दर पर ध्यान केंद्रित करना अथवा व्यक्तिगत पूर्वाग्रह अथवा टॉप-डाउन कार्यान्वयन नहीं है। यह संरक्षित व्यवहारों की प्रशंसा, प्रोत्साहन और सुदृढ़ीकरण है।
21. बीबीएस टॉप-ड्राइवन नहीं अपितु बॉटम-अप अभिगम है ।
22. बीबीएस सर्वोत्तम एवं अद्यतन संरक्षा पद्धति है जिसे यदि प्रत्येक द्वारा पूर्ण निष्ठा के साथ अपनाया जाता है तो असंरक्षित व्यवहारों में कमी और संरक्षित व्यवहारों में वृद्धि करने में मदद मिलेगी।
23. अन्ततो-गत्वा बीबीएस का उद्देश्य प्रत्येक के लिए समानुभूति और कार्य स्थल पर प्रत्येक के लिए संरक्षा का विकास करना है।
24. व्यक्ति(दृष्टिकोण, व्यक्तित्व) और पर्यावरण(प्रबंधन प्रणाली, संगठनीय वातावरण) का कार्य संपादन ही व्यवहार है। संपूर्ण संरक्षा संस्कृति में इन तीनों क्षेत्रों में ध्यान देने की आवश्यकता है, सक्रिय रूप से ध्यान देने पर इन तीनों में सुधार होता है।
25. बीबीएस व्यक्तियों द्वारा, व्यक्तियों का और व्यक्तियों के लिए है। बीबीएस मनोविज्ञान में व्यवहारवाद के स्थापित सिद्धांतों पर आधारित है।
26. बीबीएस के सिद्धांत अनुसंधान और स्थापित परिकल्पना पर आधारित है।

अवलोकन एवं फीडबैक प्रक्रिया

27. किसी कर्मचारी के पास बार-बार जाना और जब तक संरक्षित व्यवहार और स्व-अवलोकन करना और संरक्षित निष्पादन के लिए अन्य लोगों के अवलोकन की प्रक्रिया ही बीबीएस है।
28. विभाग, दिनांक, माह और समय द्वारा विशिष्ट संरक्षित और असंरक्षित व्यवहारों के अवलोकन ऑकड़ो को एकत्र करना बीबीएस है।
29. बीबीएस के प्रशिक्षित पर्यवेक्षक नियमित आधार पर संरक्षा व्यवहारों पर नजर रखते हैं।
30. व्यवहार में परिवर्तन दृष्टिकोण में परिवर्तन से पहले होता है। हम व्यवहार में परिवर्तन को महसूस करते हैं और तदनुसार हमारे दृष्टिकोण को पुनः समायोजित करते हैं।
31. बीओएफपी अर्थात् व्यवहार अवलोकन और फीडबैक प्रक्रिया बीबीएस है।
32. व्यवहार में परिवर्तन दृष्टिकोण में परिवर्तन लाता है, इसके विपरीत भी हो ऐसा आवश्यक नहीं है, इसलिए व्यवहार अवलोकन और फीडबैक प्रक्रिया पर फोकस होना चाहिए।
33. पूर्व की दुर्घटनाओं और नुकसान के रिकार्ड और ब्रेनस्टोमिंग के आधार पर क्रिटिकल व्यवहारों की चेकलिस्ट में सूचीबद्ध किया जा सकता है।
34. अलग-अलग पर्यवेक्षक भिन्न-भिन्न संरक्षित और असंरक्षित व्यवहारों पर ध्यान देंगे जिससे कि क्यों कर्मचारियों को एक-दूसरे का अवलोकन करने की आवश्यकता है।
35. अवलोकन दिनचर्या और संरक्षा प्रक्रिया में निरंतर सुधार को स्थापित करे।
36. फीडबैक ऑन-द-स्पॉट एवं ग्राफिकल हो सकते है और साप्ताहिक/मासिक रूप से संक्षिप्त विवरण दी जाए।
37. फीडबैक वास्तविक मामलों पर आधारित एक पारस्परिक क्रिया है, इसलिए फीडबैक देने और लेने में अपने स्वयं के दृष्टिकोण में कोई संदेह नहीं रखे, हमारे स्वयं के व्यवहारों की पूरी जानकारी प्रदान करता है।
38. फीडबैक तुरंत और वन-ऑन-वन दिए जाएं, विशिष्ट व्यवहारों का सुदृढ़ीकरण किया जाए और उदाहरण स्थापित करने के लिए संरक्षित व्यवहारों की प्रशंसा करें।
39. पर्यवेक्षक-पर्यवेक्षी एक जीत-जीत प्रक्रिया है जिसमें दोनों का विकास होता है।
40. हिंदी भाषा में, गंभीरता और संबद्धता के साथ देखना(अवलोकन) और बोलना(बातचीत) ही बीबीएस है।

41. बीबीएस चेकलिस्ट में कर्मचारी का नाम शामिल नहीं किया जाता है।
42. पर्यवेक्षक सह-कर्मियों के अवलोकन योग्य संरक्षित एवं असंरक्षित व्यवहारों को लक्ष्य बनाते हैं।
43. ज्ञानात्मक फीडबैक की तुलना में व्यावहारिक फीडबैक ज्यादा प्रभावी है।
44. सह-कर्मियों द्वारा स्वीकार किए जाने में अवलोकन और शिक्षण कुछ समय ले सकता है।
45. अभ्यास और चेकलिस्ट के प्रयोग से अवलोकन और फीडबैक के हुनर में सुधार होता है।
46. सप्ताह के दौरान यादृच्छिक रूप से निश्चित समय पर नहीं ऑबजरवेशन सेम्पलिंग किया जाना चाहिए।
47. अवलोकन टिप्पणी : गलत औजार का प्रयोग किया, इसके बजाय यह कहे कि क्यों और कौनसा उपकरण प्रयोग में लाने की आवश्यकता है।
48. संरक्षित और जोखिम व्यवहारों का अवलोकन करें और समस्या का समाधान, फॉलो-अप आदि के लिए विस्तृत टिप्पणियों का प्रयोग करें।
49. दो(अनुभवी पर्यवेक्षक के साथ नए पर्यवेक्षक) की टीम में अवलोकन करें।
50. पूर्ण चेकलिस्ट के अनुसार पर्यवेक्षक संरक्षित और असंरक्षित व्यवहारों पर पर्यवेक्षी को फीडबैक प्रदान करते हैं।
51. व्यवहारों में सुधार के लिए पर्यवेक्षक तुरंत फीडबैक प्रदान करते हैं।
52. संरक्षा के बारे में संरक्षा शिक्षण प्रोत्साहन से संचार में खुलापन आता है।
53. कार्य स्थल पर संरक्षा के लिए संरक्षा शिक्षण सतत अनुस्मारक के रूप में कार्य करता है।
54. प्रत्येक विभाग में संरक्षा पर्यवेक्षक/शिक्षक चेक लिस्ट और शिक्षण प्रक्रिया(समस्या, समाधान, अनुवर्ती) पर अलग-अलग क्रिटिकल व्यवहारों(पीपीई का प्रयोग, बॉडी पॉजिशनिंग उपकरणों का प्रयोग) का प्रयोग कर सकते है।
55. बीबीएस पर्यवेक्षक की कुछ विशेषताएं है : अन्यों के लिए चिंतित होना, पर्यवेक्षी की गलती को सुधारने के लिए स्वत: पहल करना, संरक्षित व्यवहार पर आपसी समझ का विकास करना, संरक्षित व्यवहार पर स्व-अवलोकन के लिए पर्यवेक्षी के स्वभाव में परिवर्तन करना।

सक्रिय रूप से ध्यान रखने से एक बीबीएस पर्यवेक्षक को क्या हासिल होता है?

सक्रिय रूप से ध्यान रखना(सुनना, प्रशंसा करना, स्वीकार करना) बीबीएस के महत्वपूर्ण पहलुओं में से एक है, प्रत्येक को यह समझने की आवश्यकता है कि सक्रिय

रूप से ध्यान रखने से एक बीबीएस पर्यवेक्षक को क्या हासिल होता है? यहां बीबीएस प्रशिक्षण के प्रतिभागियों के कुछ प्रत्युत्तर दिए गए है:

• सकारात्मक संरक्षा पद्धति	• भयहीनता
• उत्पादन में वृद्धि	• सक्रिय रूप से ध्यान रखना
• कार्य संतुष्टि	• मदद की भावना
• स्नेह एवं दुलार	• आनंद, आत्मविश्वास
• बंधन	• खुलापन
• स्वामित्व	• असंरक्षित कृत्यों में कमी, और
• टीम वर्क	• सकारात्मक प्रत्युत्तर

उपर्युक्त सभी प्रत्युत्तर स्पष्ट रूप से इंगित करते है कि संस्थान में सकारात्मक संरक्षा संस्कृति के निर्माण में बीबीएस सहयोग प्रदान करता है।

फीडबैक देखने और देने में कुछ समस्याएं :

1. पर्यवेक्षी को चेकलिस्ट सौंपना और उसे कार्य करने के लिए कहना,
2. केवल देखना, फीडबैक नहीं देना,
3. दो पर्यवेक्षी एक-दूसरे से सहमत नहीं हो सकते हैं,
4. पर्यवेक्षक किसी पर्यवेक्षी को बोलने की अनुमति नहीं दे सकते है,
5. टिप्पणी वाला कॉलम खाली रहता है,
6. केवल आलोचना, प्रशंसा नहीं है,
7. बीओएफपी के लिए अपर्याप्त समय,
8. पर्यवेक्षी हाजिर, जवाबी दिखाता है तो पर्यवेक्षक को यह स्वीकार नहीं है,
9. पर्यवेक्षक उसके अवलोकनों की गुणवत्ता पर पर्यवेक्षी से फीडबैक लेने को तैयार नहीं है।

वे कारण जिनकी वजह से एक बीबीएस पर्यवेक्षक अवलोकन और फीडबैक प्रक्रिया में असफल रह सकता है :

- प्रतिक्रिया दर्शाने का आवश्यक तरीका।
- दूसरों के सामने फीडबैक देना।
- वह स्वयं ही एसओपी का पालन नहीं कर रहा है।
- अनुपयुक्त संचार।
- फीडबैक प्रेषित करने का गलत तरीका।

- सोहार्द्रपूर्ण संबंधों का अभाव।
- पर्यवेक्षी के कार्य के बारे में जानकारी नहीं होना।
- पर्यवेक्षी को सही करने के लिए धैर्य का अभाव।
- प्रत्युत्तर क्षमताओं(गुणों) का अभाव।
- असंरक्षित व्यवहार की बजाय व्यक्ति का उल्लेख करना।
- उचित फॉलो-अप नहीं लेना।
- व्यक्तिगत दुर्भावना।
- श्रेष्ठता-मनोग्रंथि।
- यह बात संप्रेषित करने में असफल रहना कि हमारा उद्देश्य दुर्घटना युक्त संस्कृति से है।

उपर्युक्त बिंदुओं को ध्यान में रखते समय, एक बीबीबीएस पर्यवेक्षक को अवलोकन और फीडबैक प्रक्रिया पर बहुत ज्यादा औपचारिक नहीं होना चाहिए, उसे तानाशाह नहीं बनना चाहिए। एक पर्यवेक्षक को पर्याप्त रूप से लचीला और गंभीर होना चाहिए ताकि वह पर्यवेक्षी के साथ बात-चीत कर सके।

पर्यवेक्षी की फीडबैक प्रक्रिया के दौरान बीबीएस पर्यवेक्षक में निम्नलिखित अंतरवैयक्तिक क्षमताएं आवश्यक है:

- विश्वास का निर्माण करना।
- प्रत्येक के अहम का प्रबंधन करना।
- बदलते व्यवहार में धैर्य रखना।
- सकारात्मक बातचीत को सुनिश्चित करने के लिए हल्का मजाक।
- संगीत सुनना।
- बातचीत में मानवीय स्पर्श का निर्माण करना।
- उर्जस्वी उपस्थिति, सक्रिय बनना न कि निष्क्रिय।
- एक दोस्त बनना न कि बॉस।
- व्यक्तिगत तनाव से शिथिल रहना।

एक-दूसरे को प्रभावित करने वाली उपर्युक्त कार्यशाला स्व-विकास में मदद करेगी ताकि पर्यवेक्षक सक्रिय रूप से ध्यान रखते हुए प्रभावपूर्ण रूप से बीओएफपी का संचालन कर सके।

बीबीबीएस कार्यान्वयन एवं संचालन समिति

बीबीएस व्यक्तियों द्वारा, व्यक्तियों का और व्यक्तियों के लिए है। यह मनोविज्ञान में व्यवहारवाद के स्थापित सिद्धांतो पर आधारित है। प्रत्येक विभाग में व्यावहारिक संरक्षा

परियोजना का कार्यान्वयन बीबीएस पर एक दिन प्रशिक्षण को पूर्ण करने के बाद प्रारंभ होता है। बीबीएस की संकल्पना एवं प्रक्रिया को प्रत्येक स्तर के व्यक्तियों द्वारा आसानी से समझा और शीघ्र ही लागू किया जा सकता है। प्रत्येक विभाग हरेक सप्ताह और माह में अपने कामगारों के असंरक्षित और संरक्षित व्यवहारों का ग्राफ प्रस्तुत कर सकता है। इस प्रकार कोई भी माह वार असंरक्षित व्यवहारों में कमी और संरक्षित व्यवहारों में वृद्धि की तुलना कर सकता है। वर्ष की प्रत्येक तिमाही में, सारे विभाग उनके कामगारों और ऑपरेटरों के असंरक्षित और संरक्षित व्यवहारों की तुलना कर सकते है।

56. बीबीएस कार्यान्वयन में बाइ-इन वर्क फॉर्स का अभाव, स्पष्टता के साथ परिभाषित नहीं किए गए असंरक्षित व्यवहार, दुर्घटना के लिए जिम्मेदार व्यवहारों हेतु दुर्घटना रिकार्डस/बाल-बाल बचे दुर्घटनाओं का उचित रूप से विश्लेषण नहीं करने जैसी समस्याएं शामिल है।
57. बीबीएस को छोटे स्तर में आरंभ कर बड़े स्तर तक लागू किया जा सकता है जिसका तात्पर्य है कि एक विभाग से शुरूआत करे। याद रखें कि सर्वोत्तम पद्धति कोई भी नहीं है। बीबीएस को भी कस्टमाइज्ड किया जा सकता है।
58. संगठ्नात्मक प्रणाली के रूप में बीबीएस को एकीकृत किए जाने की आवश्यकता है। संरक्षा पर यह एक नई पहल, नया महत्व और नई दिशा है।
59. बीबीएस की सफलता का रहस्य यह है कि संरक्षा नियंत्रण प्रत्येक कर्मचारी के हाथों में है, वे स्वयं को जिम्मेदार और सशक्त महसूस करते है।
60. बीबीएस के कार्यान्वयन से पहले और बाद में इसके प्रभावों की तुलना करने के लिए संरक्षा जागरूकता सर्वेक्षण के माध्यम से सभी विभागों में कर्मचारियों के संरक्षा जागरूकता स्तरों की पहले और बाद की मात्रा बीबीएस में होनी चाहिए।
61. बीबीएस के सफलता की गति वर्तमान दुर्घटना/दुर्घटना दर और इसके कार्यान्वयन की तत्परता पर निर्भर करती है।
62. बीबीएस की सफलता एक मजबूत संचालन समिति पर निर्भर करती है जो कि इसके दायित्वों और जिम्मेदारियों को स्पष्ट रूप से परिभाषित करती है।
63. प्रत्येक माह विभाग संरक्षित व्यवहारों का जो उच्चतम प्रतिशत दर्शाता है उसे उचित रूप से प्रस्तुत किया जाए।
64. वरिष्ठ प्रबंधकों और फर्स्ट लाइन के नेतृत्व और व्यावहारिक संरक्षा प्रबंधन समर्थन को मात्रात्मक रूप से नापा जा सकता है।
65. कार्यान्वयन टीम/संचालन समिति नियमित रूप से बीबीएस की प्रगति पर नजर रख सकती है।

66. दीर्घ काल में, बीबीएस प्रभावी है क्योंकि यह दुर्घटनाओं को कम करता है जिनकी लागत बहुत ज्यादा होती है।
67. शुरूआत में, बीबीएस को एक या दो विभागों में कार्यान्वित किया जा सकता है और फिर दूसरे विभागों में भी लागू किया जा सकता है।
68. इसके लिए एक संचालन समिति की आवश्यकता है जो एक वरिष्ठ प्रबंधक, एक फ्रंट लाइन मैनेजर और लगभग दस बीबीएस प्रशिक्षित संरक्षा पर्यवेक्षकों से मिल कर बनी हो।
69. प्रबंधन की प्रतिबद्धता और समर्थन के साथ व्यावहारिक संरक्षा पर यह एक कर्मचारी की गतिविधि है।
70. प्रबंधन पर्यवेक्षकों को अवलोकन दौरों, फीडबैक सेशन को आयोजित करने के लिए समय और प्रदर्शन के लिए संगठित डाटा के विश्लेषण की अनुमति देता है जिसके बिना बीबीएस असफल रहेगा।
71. व्यक्तियों से यह कहा जाएगा कि वे स्वैच्छिक रूप से या तो पर्यवेक्षक या फिर संचालन समिति का हिस्सा बनें। ये व्यक्ति जिम्मेदारियां/दायित्व निभाएंगे।
72. अनुसंधान यह इंगित करता है कि बीबीएस ने अपने कार्यान्वयन से छह माह से एक वर्ष के भीतर दुर्घटना दरों में 40-75% की कमी की है।
73. भारत में, एस्सार, आईटीसी, रिलायंस, सीएफसीएल, बैयर, एचपीसीएल, ओएनजीसी, एम एंड एम, सेण्डोज, टाटा अडाणी समूह और कई अन्य निगमों ने बीबीएस का कार्यान्वयन आरंभ कर दिया है।

पर्यवेक्षकों को अपने अवलोकनों को दिशा-निर्देशित करने के लिए व्यवहारों की चेकलिस्ट की आवश्यकता होती है

पर्यवेक्षक .... दिनांक ....

विभाग/जॉब .... समय ....

व्यवहार श्रेणियां	संरक्षित	जोखिम पर	असुरक्षित स्थिति
<u>पीपीई</u> पीपीई जैसे, आई ग्लासेस, हियरिंग प्रोटेक्शन, दस्तानें, व कठोर टोपी आदि का प्रयोग			
<u>हाउसकीपिंग</u> कार्य क्षेत्र का ढंग से रख-रखाव किया			

गया है, जैसे-ट्रैश और स्क्रैप को उठा लिया गया है, फर्श पर कुछ फैला हुआ नहीं है, रास्ते में काई रूकावट नहीं है, सामग्री और उपकरणों को संगठित रूप से रखा गया है........... आदि			
<u>औजारों और उपकरणों का प्रयोग</u> कार्य के लिए सही औजारों का प्रयोग, औजारों का उचित प्रयोग और औजार सही स्थिति में हों			
<u>बॉडी की स्थिति/बचाव</u> शरीर के हिस्सों की स्थिति/बचाव जैसे-लाइन ऑफ फायर से बचना, पिंच पाइंटस से बचना............			
<u>सामग्री रखरखाव</u> जैसे –उठाने, धक्का देने और खींचते समय बॉडी मैकेनिक्स, सहायक युक्तियों का प्रयोग................			
<u>संचार</u> मौखिक और अमौखिक बातचीत जो संरक्षा को प्रभावित करे			
<u>प्रक्रियाओं का पालन</u> जैसे-परमिट प्राप्त प्राप्त करना एवं पालन करना, एसओपी लॉक आऊट, टेंग आऊट प्रक्रियाओं का पालन करना			
दृश्य पर जोर देना(सावधानी)			
कार्य करते समय मोबाइल का प्रयोग करना			

बीबीएस चेकलिस्ट तैयार करने में आने वाली कुछ समस्याएं

चेकलिस्ट तैयार करने, अवलोकन और फीडबैक देने में कुछ समस्याएं हैं। बीबीएस प्रशिक्षण में भाग लेने वाले प्रतिभागियों(संख्या-100) ने निम्नानुसार समस्याएं बताईं :

1. व्यवहार श्रेणियां उल्लिखित नहीं थी।
2. जांच बिंदुओं को व्यावहारिक रूप में जैसे कि प्रयोग करना, दूर करना उल्लिखित नहीं किया गया था।
3. कुछ व्यवहार श्रेणियों जैसे कि कार्य क्षेत्र और हाऊसकीपिंग, बॉडी मैकेनिक्स और एरगोनोमिक्स को एक साथ रखा जा सकता था।
4. जांच सूची को उचित रूप से तैयार नहीं किया गया है, कुछ व्यवहार श्रेणियों को नजर अंदाज किया गया है।

जब आप अपने संस्थान में बीबीएस को कार्यान्वित करने जा रहे है तो क्या समस्याएं आपके सामने आएगी ?

बीबीएस प्रशिक्षुओं(संख्या - 350) के प्रत्युत्तर निम्नानुसार है :

- अतितनाव : पर्यवेक्षक को यह महसूस हो सकता है कि उनका कार्यभार बढ़ गया है।
- मुझे क्या मिलेगा? पर्यवेक्षक को यह महसूस हो सकता है कि बदले में उसे क्या मिलेगा।
- क्या पर्यवेक्षी इसे सही संदर्भ में लेगा।
- बीबीएस हेतु अदला-बदली के संबंध में श्रमिकों का भय।
- बीबीएस का लाभ उठाना - दैनिक कार्य नहीं करना।
- पहले विद्यमान असंरक्षित परिस्थितियों को दूर करना/हटाना।
- बीबीएस को संस्थान में अन्य समितियों के साथ एकीकृत करना।
- बीबीएस के लिए अतिरिक्त समय/श्रम शक्ति।
- अधिक मात्रा में दस्तावेजीकरण।
- अंतर वैयक्तिक संचार और
- बीबीएस को लागू करने में आने वाली लॉचिंग समस्याएं।

बीबीएस की लॉचिंग में निम्नलिखित कार्य योजना अपेक्षित है :

1. प्रबंधन की प्रतिबद्धता।
2. पूरे संयंत्र में कर्मचारियों का प्रशिक्षण।
3. बीबीएस प्रशिक्षित कर्मचारियों से चयनित कर पर्यवेक्षक तैयार करना।
4. बीबीएस प्रशिक्षित कर्मचारियों में से 8-10 व्यक्तियों की संचालन समिति का गठन करना।

संचालन समिति की जिम्मेदारियां :

5. जांच सूचियां तैयार करना।
6. व्यवहार अवलोकन और फीडबैक प्रक्रिया के लिए क्षेत्रों/विभागों का चयन करना।
7. समय सूची तैयार करना।
8. दस्तावेजीकरण, विश्लेषण और प्रदर्शन।
9. संस्थान में बीबीएस की प्रगति को निर्धारित करने की समीक्षा करना।
10. बीबीएस अवलोकनों और बैठकों आदि के लिए समय आबंटित करना।
11. बीबीएस की लांचिंग के लिए अंतिम तिथि तय करना।

बीबीएस प्रशिक्षण प्रतिभागियों से दस अपेक्षाएं

- हम दुर्घटना बचाव में नं. 1 बनना चाहते है।
- हम कैसे मानवीय व्यवहार को सही करें?
- कैसे व्यवहार संरक्षा में महत्वपूर्ण है?
- हम कैसे प्रवृत्ति और व्यवहार को बदल सकते हैं?
- हम कैसे संरक्षित व्यवहार का विश्लेषण करें?
- बीबीएस का कार्यान्वयन कैसे करे?
- कैसे असंरक्षित व्यवहार को ढूँढे।
- कैसे श्रमिकों को संरक्षा के बारे में समझाएं?
- क्या संरक्षित व्यवहारों को मापा जा सकता है?
- 'संरक्षित व्यवहार' को कैसे बनाए रखा जाए, यह एक चुनौती है ?

बीबीएस पद्धति में प्रशिक्षण प्रतिभागियों के लिए प्रवृत्तियां एवं योग्यताएं:

निम्नलिखित अभ्यास, अभिज्ञता और चर्चा बीबीएस पद्धति के लिए आवश्यक प्रवृत्तियों और योग्यताओं के निर्माण में सहायक होगी :

1. संयंत्र संरक्षा गतिविधियों को पुन: गिनना।
2. संस्थान में या जॉब में संरक्षा के स्तर के साथ आप कैसे संतुष्ट हैं।
3. आपके संस्थान का दुर्घटना इतिहास और संरक्षा पर प्रबंधन का हस्तक्षेप।
4. आपके कार्य क्षेत्र में किसी असंरक्षित व्यवहार को याद रखना।
5. किसी भी स्पष्टीकरण के लिए कार्यशाला पाठ्य सामग्री को सक्रिय रूप से देखना।
6. बीबीएस संकल्पना पर प्रश्न और उत्तर सत्र।
7. बीबीएस के संदर्भ में प्रवृत्ति-व्यवहार संबंध।

8. क्यों एक बीबीएस पर्यवेक्षक अवलोकन और फीडबैक प्रक्रिया में असफल हो सकता है।
9. सक्रिय रूप से ध्यान रखने से एक बीबीएस पर्यवेक्षक को क्या प्राप्त होगा।
10. बीबीएस पर्यवेक्षक की विशेषताएं।
11. अवलोकन और फीडबैक देने में समस्याएं।
12. जांच सूची तैयार करने में व्यावहारिक होना।
13. बीओएफपी में व्यावहारिक होना।
14. बीबीएस कार्यान्वयन/संचालन समिति पर चर्चा।
15. बीबीएस के कार्यान्वयन में प्रतिभागियों को क्या समस्याएं आती है।
16. बीबीएस को लॉंच करने में क्या योजना अपनाए जाने की आवश्यकता है।
17. बीबीएस कार्यान्वयन टीम का गठन करना और कार्य योजना तैयार करना।
18. बीबीएस याहू ग्रुप से जुड़ना : BBSINDIA@yahoogroop.com

बीबीएस प्रशिक्षण प्रतिभागियों(संख्या – 350) के कुछ प्रत्युत्तर :

- बीबीएस श्रमिकों के बीच बातचीत को बढ़ावा देगा।
- संरक्षा प्रोफेशनल के रूप में इसने मेरी जानकारी को बढ़ाया है – एक संरक्षा अधिकारी।
- हम कार्यक्षेत्र पर दैनंदिन आधार पर बीबीएस का कार्यान्वयन कर सकते है।
- अधिक से अधिक कर्मचारियों को बीबीएस का प्रशिक्षण लेना चाहिए।
- सभी कर्मचारियों को शामिल करना लाभदायक होगा।
- बीबीएस का प्रशिक्षण प्रतिभागियों को एक लघु सलाहकार बनाता है।
- वरिष्ठ प्रबंधकों के लिए एक दिन का समालोचना प्रशिक्षण कार्यक्रम बीबीएस के कार्यान्वयन में मदद करेगा।

असंरक्षित और संरक्षित व्यवहारों का विश्लेषण :

नीचे दी गई सारणी में इंगित किए गए 5 संस्थानों के 137 कर्मचारियों को घर, कार्यक्षेत्र और घर से कार्यालय तक जाने व आने के रास्ते पर दैनंदिन आधार पर शामिल उनके संरक्षित और असंरक्षित व्यवहारों को रिपोर्ट करने के लिए कहा गया था। विश्लेषण उनके असंरक्षित व्यवहारों का औसत 52.6% और उनके संरक्षित व्यवहारों का औसत 47.4% दर्शाता है।

सारणी : असंरक्षित और संरक्षित व्यवहार

संस्थान	कर्मचारियों की संख्या	असंरक्षित व्यवहारों की संख्या एवं प्रतिशत	संरक्षित व्यवहारों की संख्या एवं प्रतिशत	व्यवहारों की कुल संख्या
1	30	311 48%	330 52%	643
2	28	230 49%	239 51%	469
3	23	158 54%	135 46%	291
4	31	219 47%	243 53%	462
5	25	226 65%	124 35%	350
कुल	137	औसत 52.6%	औसत 47.4%	

विश्लेषण यह बताता है कि प्रत्येक कर्मचारी जब वह संरक्षित व्यवहार कर रहा होता है तब वह समानान्तर रूप से घर, कार्यक्षेत्र और घर से कार्यालय आते व जाते समय सड़क पर आधे से अधिक बार असंरक्षित व्यवहार भी कर रहा होता है। इसलिए उनकी संरक्षा के लिए उनको किसी और की आवश्यकता होती है जिसे बीबीएस के सिद्धांतो में बताया गया है कि व्यक्तियों को प्रत्येक की संरक्षा अर्थात पीयर-टू-पीयर संरक्षा का ध्यान रखने की आवश्यकता है।

दक्षिण भारत की एक दवा कंपनी का उदाहरण (दिसंबर 2010)

व्यावहारिक संरक्षा पर दो दिन तक 108 प्रतिभागियों से विचार-विमर्श किया गया और उन्होंने निम्नलिखित मुद्दों पर अपने विचार व्यक्त किए :

1. जब व्यक्ति ऐसी पृष्ठभूमि/शहरों से आते हैं जहां जोखिम लेना सामान्य बात है, तो कैसे हम एक संस्थान में संरक्षित व्यवहार कर सकते हैं ?
2. एक विभागाध्यक्ष का कहना है कि 'व्यावहारिक संरक्षा वाहन चलाने, रसोईघर या फैक्टरी में काम करने के संबंध में जीने का एक तरीका है।
3. बीबीएस एक से सबको, जैसा कि संरक्षा अधिकारी करते है के बजाय एक के बाद एक को प्रशिक्षित कर रहा है। बीबीएस को कार्यान्वित करने के बाद हमारे संयंत्र में छह माह के भीतर दुर्घटनाओं की संख्या 102 से घटकर 32 रह गई है। बीबीएस नए अवलोकनों को देखकर शॉप-फ्लोर पर दिन नए मित्र बनाता है।
4. उन्होंने बताया कि संयंत्र में संरक्षित रहने का अर्थ है कि हर समय मुस्तैद रहना। कोई भी बाहरी मुस्तैदी व्यवहार में परिवर्तन उत्पन्न करती है। यदि

हम व्यावहारिक रूप से सही है तो संरक्षा इससे जुड़ी हुई है। श्रमिकों को उनकी क्षेत्रीय भाषाओं में शिक्षित करे।

व्यवहार आधारित संरक्षा पर एक भारतीय अनुभव

अब तक बीबीएस ने पिछले 15 सालों में भारत में विभिन्न क्षेत्रों में लगभग 200 बीबीएस कार्यशालाएं आयोजित की हैं। प्रत्येक संस्थान ने बीबीएस को अपनी सांस्कृतिक पृष्ठभूमि के अनुसार लागू और समायोजित किया है। कुछ संस्थानों ने पहले वरिष्ठ प्रबंधन को बीबीएस संकल्पना, प्रक्रिया और कार्यान्वयन के बारे में बताया और फिर नीचे के स्तर को बताया। अन्यों ने इसे निचले स्तर के कर्मचारियों और संविदा श्रमिकों से प्रारंभ किया। जबकि कुछ अन्य संस्थानों ने उनके मध्यम प्रबंधन और गैर-प्रबंधन कर्मचारियों को साथ में रखकर इसे प्रारंभ किया। कुछ स्थानो पर, संस्थानों ने 'प्रशिक्षकों को प्रशिक्षण' कार्यक्रम आयोजित किया और फिर इन इन-हाउस प्रशिक्षकों ने विभिन्न विभागों के अन्य कर्मचारियों को प्रशिक्षित किया। व्यावहारिक रूप से संस्थानों में एक दिन की बीबीएस कार्यशाला प्रतिभागियों को संकल्पना, प्रक्रिया और कार्यान्वयन को समझने में मदद करती है। अगला कदम बीबीएस पर्यवेक्षकों को चिह्नित करने को होना चाहिए और फिर उन्हें व्यवहार अवलोकन फीडबैक प्रक्रिया पर विस्तृत रूप से प्रशिक्षित करें। व्यावहारिक रूप से प्रत्येक विभाग में 20% कर्मचारी बीबीएस पर्यवेक्षक के रूप में प्रशिक्षित होने चाहिए। एक महीने बाद मुख्य प्रशिक्षक द्वारा एक दिन की समीक्षा प्रक्रिया को परिष्कृत करने में मदद करती है। उत्साहपूर्ण रूप से बीबीएस शीघ्र ही परिणाम देना आरंभ कर देता है। इन-हाउस बीबीएस मुख्य समिति का निर्माण करना बीबीएस गतिविधि और प्रक्रिया के समन्वय में सहायक होता है।

भारतीय संस्थानों में संरक्षा संस्कृति का निर्माण करने, संरक्षित व्यवहारों को प्रोत्साहित करने और असंरक्षित व्यवहारों में कमी करने के संबंध में बीबीएस ने सकारात्मक परिणाम दिए है। बीबीएस ने पहले से एक बेहतर दुर्घटना बचाव अभ्यास प्रदान किया है। कर्मचारियों को बीबीएस की जानकारी इस तथ्य को समझने के लिए एक समृद्ध और नवीन अनुभव है कि कार्यस्थलों पर बाल-बाल बचे दुर्घटनाओं को रोकने के क्रम में हमे बीबीएस पद्धति के माध्यम से पहले असंरक्षित/जोखिम व्यवहारों से निपटने की आवश्यकता है। संरक्षा अनुपालन के रूप में ओएचएसएएस 18001:2009 में व्यावहारिक पहलुओं को शामिल करने से, भारतीय संस्थानों ने अपने प्रशिक्षण अनुप्रयोगों में बीबीएस को गंभीरता से लिया है।

बीबीएस परियोजना के लिए पूर्व-आवश्यकताएं :

1. संस्थान का(पिछले 5 वर्षों की) एक दुर्घटना इतिहास,
2. विद्यमान संस्था उपायों की सूची
3. संस्थान की रूपरेखा,
4. वर्तमान दुर्घटना/दुर्घटना दर,
5. पिछले एक साल में रिपोर्ट की गई बाल-बाल बची दुर्घटनाएं,
6. व्यवहार आधारित संरक्षा प्रक्रिया को लागू करने में आपकी तत्परता,
7. संस्थान का संरक्षा वातावरण सर्वेक्षण

किसी भी संरक्षा सर्वेक्षण का उद्देश्य एक कंपनी को यह सूचित करना होता है कि व्यावहारिक संरक्षा पद्धति को सफलतापूर्वक प्रारंभ किए जाने में संभवतः पहले से ही विद्यमान कोई संरक्षा समस्या रूकावट पैदा कर सकती है। शैक्षणिक रूप से व्युत्पन्न संरक्षा वातावरण सर्वेक्षण और/अथवा संकेंद्रित समूह चर्चाओं सहित असंख्य प्रकार के तैयार सर्वेक्षण विद्यमान है।

बीबीएस को लाँच करने से पूर्व प्रबंधन द्वारा पूछे जाने वाले प्रश्न:

1. पर्यवेक्षकों का चयन कैसे किया जाए ?
2. बीबीएस को कैसे लाँच किया जाए ?
3. क्या कोई कार्यान्वयन समस्याएं है ?
4. कैसे और किसे संचालन समिति का गठन करना चाहिए ?
5. क्या अंतरवैयक्तिक योग्याताओं पर पर्यवेक्षकों के लिए पृथक प्रशिक्षण आवश्यक है ?
6. क्या संचालन समिति को कर्तव्यों और जिम्मेदारियों के लिए पृथक प्रशिक्षण की आवश्यकता है ?
7. क्या एक पर्यवेक्षक को उसके सीने पर बैज की आवश्यकता है ?
8. बीबीएस परियोजना के लिए क्या शीर्षक होना चाहिए ?
9. कौन पर्यवेक्षक होना चाहिए ?

बीबीएस परियोजना के लिए कुछ नाम :

1. व्यावहारिक संरक्षा परियोजना(बीएसपी)
2. सर्वोत्तम संरक्षा परियोजना(बीएसपी)
3. आई से आई संरक्षा(बीबीएस)
4. पीयर टू पीयर सेफ्टी(पीपीएस)

5. भाई-बंधु सुरक्षा(बीबीएस)
6. सर्वोत्तम सुरक्षा मार्ग(बीएसएम)
7. देखो बोलो सुरक्षा(डीबीएस परियोजना)
8. असंरक्षित से संरक्षित परियोजना(यूएसपी)
9. हर एक के लिए संरक्षा परियोजना

निष्कर्ष :

संस्थानों में अभियांत्रिकी समाधानों के संबंध में एक अच्छी सीमा तक उपलब्धि प्राप्त कर ली गई है लेकिन संरक्षा प्रबंधन में व्यावहारिक अभियांत्रिकी को सीखना एक बहुत बड़ी चुनौती है।

दुर्घटनाएं कम करने के लिए, प्रबंधन ने कई संरक्षा हस्तक्षेप किए है जैसे : जोखिम निर्धारण, सुझाव योजना, प्रशिक्षण, संरक्षा समिति, अंकेक्षण, प्रेरणादायक कार्यक्रम(प्रश्नोत्तरी, पुरस्कार, प्रोत्साहन) एसओपी, संयंत्र निरीक्षण, कार्य परमिट पद्धति आदि। इन संरक्षा प्रबंधन पद्धतियों में से अधिकतर का उद्देश्य असंरक्षित परिस्थितियों को नियंत्रित करना है, जब कि 80-95% दुर्घटनाएं असंरक्षित कृत्यों और व्यवहारों के कारण होती है।

व्यक्ति असंरक्षित व्यवहार करते है क्योंकि इससे उनका समय और प्रयास(शोर्ट कट लेना अथवा पीपीई का प्रयोग नहीं करना) की बचत होती हैं। पर्यावरणीय समाधान प्रभावी रूप से कार्य नहीं करते है क्योंकि व्यक्ति रक्षा कवच को हटा देते है और अनुचित हाउसकीपिंग में कार्य करते है। दंड का प्रभाव सकारात्मक या नकारात्मक हो सकता है। व्यवहार परिवर्तन हमारा सहायता नहीं करता है क्योंकि यह वास्तविक रूप से व्यवहार में परिवर्तन नहीं करता है। इसलिए बेहतर संरक्षा परिणामों के लिए बीबीएस का प्रयोग किया जा सकता है।

बीबीएस प्रेरणा प्रतिक्रिया सिद्धांत पर आधारित है जो यह मानकर चलता है कि असंरक्षित व्यवहार को परिवर्तित किया जा सकता है और संरक्षित व्यवहार को बढ़ाने के लिए बाहरी प्रेरणा की बारंबार आवश्यकता होती है जो कि बीओएफपी के माध्यम से प्रदान की जाती है।

बीबीएस इस बात पर बल देता है कि हर एक कि संरक्षा के लिए सक्रिय रूप से जिम्मेदारी ले। अवलोकन योग्य व्यवहार अपेक्षित परिणाम पर जोरे देने का लक्ष्य बनाएं अर्थात असंरक्षित से संरक्षित व्यवहार में परिवर्तन, वर्षों के दौरान विभागों में संरक्षित और जोखिम व्यवहारों के प्रतिशत को समझने के लिए प्रत्येक व्यक्ति अथवा समूह के प्रतिदिन/साप्ताहिक/मासिक व्यवहारों के रूझानों पर निगरानी रखे।

औद्योगिक संरक्षा की ओर बीबीएस एक गुणात्मक पद्धति है जोकि संरक्षा के संबंध में कर्मचारियों की प्रवृत्ति में गुणात्मक परिवर्तन करती है। यह संस्थान में संरक्षा कार्यान्वयकों की एक सेना तैयार करती है।

किसी मामले में 50-60 प्रतिशत कर्मचारियों द्वारा पीपीई का प्रयोग नहीं किया जा रहा है यह एक सिस्टम में कमी है क्योंकि यह पीपीई के प्रयोग नहीं करने के उल्लंघन के विरूद्ध कोई कार्रवाई नहीं करता है। संपूर्ण संरक्षा संस्कृति यह सम्मिलित करता है कि संरक्षा तंत्र अपने स्थान पर है और सक्रिय है और तब बीबीएस को कार्यान्वित करने से बहुत ही उत्कृष्ट परिणाम प्राप्त होते है। एक वरिष्ठ संरक्षा प्रोफेशनल के अनुसार 'संस्थानों में संरक्षा के लिए परिणामों को बनाए रखने के लिए दंड कभी भी प्रभावी नहीं होता है।' अन्य संरक्षा प्रोफेशनल ने कहा कि, 'आने वाले वर्षों में संरक्षा के श्रेष्ठ घटकों में से बीबीएस भी एक घटक होगा।' अन्य वरिष्ठ संरक्षा प्रोफेशनल के अनुसार, 'कार्यस्थल पर आपके प्रचालनात्मक नियंत्रण हो सकते है, आपने कर्मचारियों को संरक्षा के लिए कहा होगा फिर भी मनुष्य असंरक्षित व्यवहारों के कारण दुर्घटनाओं के शिकार होते रहते है।' बीबीएस इस बात में विश्वास करता है कि हर एक का बार-बार और सक्रिय रूप से ध्यान रखने से मनोवैज्ञानिक परिवर्तन प्राप्त किया जा सकता है। बीबीएस इस बात पर जोर देता है कि जब 80-90 प्रतिशत दुर्घटनाएं और चोट असंरक्षित व्यवहारों के कारण होती है तो असंरक्षित स्थितियों की तुलना में असंरक्षित और संरक्षित व्यवहारों पर ज्यादा ध्यान केंद्रित किया जाए। व्यवहार वस्तुनिष्ठ, परिभाष्य, अवलोकनीय, सुधारात्मक और मापने योग्य है।

बीबीएस पद्धति को पूरी शक्ति के साथ लॉच करने के लिए एक प्रत्यक्ष उपस्थिति और एक स्वच्छग्राही प्रबंधन और नीचे की तरफ खुले संचार की आवश्यकता होती है, ऐसा नहीं होने पर सफल होना मुश्किल है।

प्रबंधन को अपने संस्थान में संरक्षा की ओर उनकी प्रवृत्ति का मूल्यांकन करने की आवश्यकता है। एक संस्थान के प्रचालन प्रमुख का कहना है कि 'सामान्यतः जब कार्य का भार ज्यादा होता है, हम संरक्षा को सेकंडरी समझते हैं, अब जबकि कार्य का भार कम है, हम संरक्षा को प्राथमिकता दे रहे हैं और कर्मचारियों को प्रशिक्षण दे रहे हैं।' बीबीएस को लॉच करते समय, प्रबंधन संरक्षा स्वामित्व की जिम्मेदारी श्रमिकों को सौंपता है लेकिन संरक्षा जिम्मेदारी को श्रमिकों पर थोपता नहीं है। उन्हें कर्मचारियों को नियमित रूप से परामर्श, निगरानी और प्रोत्साहित करना होगा।

बीबीएस कार्यशाला - जानकारी का आकलन प्रश्नोत्तरी

1. बीबीएस क्या है ? इसे परिभाषित करें ?
2. बीओएफपी क्या है ?
3. डीओ - आई टी क्या हैं ?
4. बीबीएस के कार्यान्वयन संबंधी दो समस्याएं क्या हैं ?
5. बीबीएस के दो भ्रम क्या हैं ?

निम्नलिखित में सही अथवा गलत पर निशान लगाएं ।

5. बीबीएस के दो भ्रम क्या हैं ?
6. विशेष व्यवहारों पर दिनांक, माह और समय वार अवलोकन डाटा एकत्र करना बीबीएस है ।

   सही/गलत
7. स्वयं और दूसरों के लिए संरक्षा ही बीबीएस है।

   सही/गलत
8. बीबीएस उच्च प्रबंधन की बजाय कर्मचारियों द्वारा चलाए जाने की पद्धति है ।

   सही/गलत
9. सभी कार्यक्षेत्रों और विभागों में ब्रीफिंग सत्र के साथ बीबीएस प्रारंभ होता है ।

   सही/गलत
10. बीबीएस इस बात में विश्वास करता है कि प्रत्येक कर्मचारी संस्थान की संरक्षा में परिवर्तन ला सकता है ।

    सही/गलत
11. बीबीएस इस बात में विश्वास करता है कि जो कुछ भी सोचा गया है, उसे पूरा किया जाए ।

    सही/गलत
12. बीबीएस आंकड़ों के आधार पर निर्णय लेने की प्रक्रिया है।

    सही/गलत
13. बीबीएस टीम वर्क है, यह ग्रुप डायनेमिक्स है।

    सही/गलत

14. बीबीएस का उद्देश्य नियमों को लागू करना, परिवर्तन के लिए दबाव बनाना, दूसरों के लिए गपशप करना, बॉस को रिपोर्ट करना नहीं है ।

सही/गलत

15. बीबीएस का उद्देश्य संरक्षित और जोखिम पूर्ण व्यवहारों एवं दुर्घटना के अवसरों की पहचान करना, जोखिम का संप्रेषण और संरक्षित समाधानों की पहचान करने में मदद करना है।

सही/गलत

16. बीबीएस पूरी कंपनी में व्यक्तियों द्वारा चलाया जाने वाला है ।

सही/गलत

17. व्यवहार उद्देश्यपरक, परिभाष्य, अवलोकन योग्य और मापने योग्य है ।

सही/गलत

18. कर्मचारी व्यावहारिक परिवर्तन हेतु विशेषज्ञ के स्रोत हैं।

सही/गलत

19. उदाहरण स्थापित करने के लिए एक-दूसरे को तुरंत फीडबैक दिया जाना चाहिए, विशेष व्यवहारों को सुदृढ़ और संरक्षित व्यवहारों की प्रशंसा की जानी चाहिए ।

सही/गलत

20. क्रिटिकल व्यवहारों पर फोकस हेतु अवलोकन के लिए चेकलिस्ट का प्रयोग करें।

सही/गलत

बीबीएस पोस्टरों एवं बैनरों के लिए सामग्री:

1. आइए हम सब मिलकर क्षतिरहित संस्कृति का विकास करें।
2. प्रत्येक कर्मचारी संगठनात्मक संरक्षा में बदलाव ला सकता है।
3. संरक्षा जनता से संचालित है।
4. असुरिक्षत एवं सुरक्षित व्यवहार की सक्रिय जिम्मेदारी बेहतर संरक्षा की प्रेरक है।
5. व्यवहार आधारित संरक्षा 'एक-दूसरे की संरक्षा' है।
6. संरक्षा सुरक्षित व्यवहारों की 'पीअर-टू-पीअर' जानकारी है।
7. एक-दूसरे की संरक्षा के प्रति दिलचस्पी पैदा करना ही संरक्षा है।
8. संरक्षा भावुकता एवं दिलचस्पी के साथ 'देखो' और 'बोलो' है।

9. सुरक्षित एवं जोखिम भरे व्यवहार को ध्यान से देखना है।
10. संरक्षा पालक सुरक्षित एवं असुरक्षित व्यवहार पर फीडबैक देते हैं।
11. बीबीएस कर्मचारियों की भावनाओं के विषय में है।
12. संरक्षा बिना दंड के अर्जित की जाती है।
13. किसी दुर्घटना के बिना व्यक्ति को बचाना।
14. सुरक्षित एवं मैत्रीपूर्ण वातावरण बनाना।
15. यदि हम सुरक्षित तरीके से काम करेंगे, तो मेरा परिवार सुरक्षित रहेगा।
16. मानव बोध के द्वारा संरक्षा।
17. बिना किसी प्रयास के 100 प्रतिशत संरक्षा।
18. बिना किसी जोखिम के 'जोखिम-भरे' व्यवहार की पहचान।
19. बीबीएस के कार्यान्वयन के बाद कर्मचारी के व्यवहार में स्वयं परिवर्तन।
20. बीबीएस मर्मस्पर्शी एवं उत्साहपूर्ण संरक्षा है।
21. प्रत्येक कर्मचारी संगठनात्मक संरक्षा में परिवर्तन ला सकता है।
22. सुरक्षित कार्यसाधक वातावरण बनाता है।
23. बीबीएस विकास एवं अच्छी संस्कृति प्रदान करता है।
24. नवागतों के लिए संरक्षा का अच्छा मार्ग।
25. बीबीएस के पालन से कोई दुर्घटना और क्षति नहीं होती।
26. बीबीएस के साथ कर्मचारियों के अहंभाव को कम किया जाता है।
27. कामगारों के बीच आपसी समझ अर्जित की जा सकती है।
28. बीबीएस जागरूकता से चौकसी तक की संरक्षा है।

बीबीएस प्रशिक्षित पर्यवेक्षकों को अनुभवात्मक जानकारी :

दिसंबर 2010 में मुंबई के एक ऑटोमोबाइल संगठन में 20 मध्य एवं वरिष्ठ प्रबंधकों को बीबीएस पर प्रशिक्षण दिया गया। उन्होंने बताया कि प्रत्येक कार्य क्षेत्र में बीबीएस टीम बनाने से कार्यस्थलों पर बीबीएस का कार्यान्वयन आसान तरीके से किया जा सकता है। प्रत्येक कार्य क्षेत्र अपनी टीम की चर्चा के अनुसार बीबीएस का कार्यान्वयन कर पाया। पहले वे असुरक्षित व्यवहार को देखते और टालते थे, अब वे सुव्यवस्थित

रूप से ऐसे व्यवहारों पर ध्यान देते हैं तथा मानवीय संबंध के साथ पर्यवेक्षितों को फीडबैक उपलब्ध कराते हैं। वे प्रत्येक तीन महीनों में बीबीएस दिवस मनाते हैं। उन्होंने सीखा कि:

1. बीबीएस हमें संरक्षा जागरूकता से व्यक्तिगत चेतावनी तक ले जाता है।
2. हमारा लक्ष्य शून्य दुर्घटना से शून्य असुरक्षित व्यवहार में परिवर्तित हो गया है।
3. यह एक सहकर्मी से दूसरे सहकर्मी की संरक्षा है।
4. यह एक जीवनरक्षक प्रक्रिया है।
5. वार्तालाप में मानवीय संबंध होने से चमत्कार हो सकता है, इससे संरक्षा के प्रति मनोवृत्ति में परिवर्तन होगा।
6. बीबीएस का विस्तार हमारे विक्रेताओं तक होना चाहिए, जो कि हमें कच्चा माल उपलब्ध कराते हैं।
7. बीबीएस पर प्रशिक्षण के बाद लोग लापरवाह या उदासीन होने की बजाय दिलचस्पी लेते हैं।
8. प्रत्येक क्षेत्र में बीबीएस की टीम बीबीएस की प्रक्रिया एवं संकल्पना के बारे में अपने कामगारों को शिक्षित करती है।
9. उनके कुल 2000 कार्यबल के प्लांट में 14000 असुरक्षित व्यवहार हैं, जिन्हें दैनिक आधार पर बीओएफपी के प्रयोग से तथा अधिक-से-अधिक पर्यवेक्षकों को भी प्रशिक्षण देकर कम करने की योजना है।

संगठन में बीबीएस संस्कृति बनाए रखने हेतु संसूचक :

1. निर्धारित फॉर्मेट में प्रशिक्षित पर्यवेक्षकों से प्राप्त बीबीएस डाटा को दैनिक रूप से मॉनीटर करना।
2. प्रतिमाह आयोजित की जाने वाली बैठक की समीक्षा करना।
3. बीबीएस प्रशिक्षित व्यक्तियों की क्रोड समिति का गठन करना।
4. संयंत्र स्थल पर बोर्ड में आठ व्यवहारों के बीबीएस बैनरों का प्रदर्शन करना।
5. सभी कामगारों/कर्मचारियों का स्थिति-वार प्रशिक्षण करना।
6. क्रोड समिति द्वारा प्रत्येक माह प्रबंधन को रिपोर्ट करना।
7. केवल प्रशिक्षित श्रमिकों को ही पर्यवेक्षण करने की अनुमति।
8. "व्यवहार प्रतिदर्शी जांचसूची" के सुरक्षित एवं असुरक्षित व्यवहारों के साप्ताहिक/मासिक प्रतिशत तथा ग्राफों का प्रदर्शन एवं विश्लेषण करना।
9. तिमाही बीबीएस प्रगति लेखापरीक्षा।

10. ईएचएस पॉलिसी/फ्रेमवर्क के अंतर्गत बीबीएस को शामिल करना।
11. असुरक्षित परिस्थितियों को सुधारने हेतु कार्य योजना।
12. प्रति सप्ताह/माह में प्रत्येक पर्यवेक्षक, प्रत्येक व्यवहार तथा असुरक्षित एवं सुरक्षित व्यवहारों के कुल योग के बीबीएस डाटा का विश्लेषण किया जाए। डाटा को कंप्यूटर पर रोजाना दर्ज किया जाए। क्रोड समिति सदस्यों द्वारा प्रत्येक पर्यवेक्षक को हर रोज सुबह व्यवहार प्रतिदर्शी का रिक्त फॉर्म वितरित किया जाए तथा शाम को उनसे भरा हुआ फॉर्म एकत्र किया जाए।
13. क्रोड़ समिति के प्रत्येक सदस्य के पास परिशिष्ट-ए(व्यवहार प्रतिदर्शी फॉर्म), परिशिष्ट-बी(बीबीएस सार सांख्यिकी फॉर्म) तथा परिशिष्ट-सी(बीबीएस संस्कृति को कैसे बनाए रखे) होने चाहिए।
14. बीबीएस का मुख्य उद्देश्य असुरक्षित व्यवहारों में कमी लाना तथा सुरक्षित व्यवहारों में बढोतरी को प्राप्त करना है।
15. इसमें शामिल प्रत्येक व्यक्ति से सकारात्मक मनोवृत्ति तथा एकनिष्ठ दृष्टिकोण सुनिश्चित करना।
16. 'प्रशिक्षक को प्रशिक्षण' कार्यक्रम से परिचय कराना। इस प्रकार, प्रत्येक विभाग से बीबीएस प्रशिक्षित सदस्य अपने संबंधित विभाग में सभी सदस्यों को प्रशिक्षित करने की जिम्मेदारी लेते हैं। प्रत्येक बीबीएस प्रशिक्षित सदस्य प्रतिमाह विभाग में से एक सदस्य को प्रशिक्षित करेगा। इस तरह, संयंत्र में बीबीएस प्रशिक्षित पर्यवेक्षकों की संख्या बढ़ेगी। बीबीएस पर्यवेक्षक आईडी बैच भी जारी किए जाएं। 25 प्रशिक्षित पर्यवेक्षक शुरूआती तौर पर 8 महीनों के भीतर 3200 व्यक्तियों को प्रशिक्षण दे सकते हैं अर्थात 25>50>100>200>400>800>1600>3200.
17. संगठन में बीबीएस का औपचारिक विमोचन(किक-ऑफ)।

प्रेक्षकों की उत्कृष्टता का परीक्षण
निम्नलिखित दस पहलुओं पर 1-10 के क्रम में स्वमूल्यांकन :
1. प्रेक्षण की गुणवत्ता
2. बीओएफपी से आपकी संतुष्टि
3. जांचसूची को भरना/आपकी समझ
4. पर्यवेक्षित के साथ परस्पर संवाद
5. आपका निश्चिंत संप्रेषण स्तर
6. कंप्यूटर डाटा प्रविष्टि कौशल

7. अंतरवैयक्तिक कौशल स्तर(फीडबैक कौशल)
8. संरक्षा में आपकी अभिरुचि
9. बीबीएस के बारे में आपकी समझ
10. पर्यवेक्षक के रूप में विकास एवं वृद्धि हेतु आपकी समझ

प्रेक्षकों के पांच 'प' व्यवहार
1. प्रश्न: प्रेक्षित से पूछे कि क्या वह सुरक्षित है।
2. प्रशंसा: उनके सुरक्षित व्यवहारों को बढ़ाने के लिए उनकी प्रशंसा करें।
3. परिवर्तन: उनके असुरक्षित व्यवहारों को सुरक्षित में बदलिए।
4. प्रशिक्षण: उनके असुरक्षित व्यवहारों के लिए शिक्षित/पुन: प्रशिक्षित करें।
5. प्रतिज्ञा: उन्हें शपथ दिलवाएं कि वे असुरक्षित व्यवहारों को नहीं दोहराएंगे।

'भाई-बंधु की सुरक्षा बीबीएस' है।

सुरक्षित एवं असुरक्षित व्यवहारों तथा परिस्थितियों का सारांश : एक मामला

दिनांक	पर्यवेक्षकों की संख्या/ दौरा किए गए क्षेत्र	सुरक्षित व्यवहार की संख्या पहली पारी	सुरक्षित व्यवहार की संख्या दूसरी पारी	असुरक्षित व्यवहार की संख्या पहली पारी	असुरक्षित व्यवहार की संख्या दूसरी पारी	असुरक्षित परिस्थिति की संख्या पहली पारी	असुरक्षित परिस्थिति की संख्या दूसरी पारी
25.9.2007	29 -स्लेट क्षेत्र - पीएफ डब्ल्यू/ई - टीएलडी	107	103	59	33	41	31
26.9.2007	30 - स्लेट क्षेत्र - पीएफ डब्ल्यू/ई - टीएलडी	141	155	49	27	47	23
27.9.2007	33 - स्लेट क्षेत्र - पीएफ डब्ल्यू/ई - टीएलडी	217	235	47	29	03	01

28.9.2007	36 - स्लेट क्षेत्र - पीएफ डब्ल्यू/ई - टीएलडी	243	274	37	14	01	01

अवलोकन:

1. सुरक्षित व्यवहारों की संख्या प्रतिदिन/पारी में बढ़ रही है।
2. असुरक्षित व्यवहारों की संख्या प्रतिदिन/पारी में कम हो रही है।
3. असुरक्षित परिस्थितियों की संख्या प्रतिदिन/पारी में कम हो रही है।

सुजलॉन कंपनी का एक केस अध्ययन

व्यवहार-आधारित संरक्षा(बीबीएस) पर एक दिवसीय कार्यशाला के अवसर पर पहले दिन 20 तकनीशियनों तथा दूसरे दिन 13 इंजीनियरों को प्रशिक्षित किया गया। उन्हें स्वयं द्वारा तथा सहकर्मियों द्वारा नियमित रूप से किए जा रहे असुरक्षित व्यवहारों को दर्शाने के लिए कहा गया, जिसका परिणाम निम्नानुसार है :

व्यक्तिगत असुरक्षित व्यवहारों की संख्या	सहकर्मियों के असुरक्षित व्यवहारों की संख्या
तकनीशियन: 198/20 – औसतन 10	74/20 – औसतन 4
इंजीनियर : 88/13 – औसतन 7	61/12 – औसतन 5

यह दर्शाता है कि असुरक्षित व्यवहार व्यक्तिगत सहकर्मी द्वारा अधिकृत होते हैं, जो कि किसी भी बाल-बाल बचे दुर्घटना या चोट लगने का मूल कारण है।

प्रशिक्षण के दौरान, 20 तकनीशियनों तथा 13 इंजीनियरों को कार्यस्थल पर अपने सह-कर्मियों के सुरक्षित एवं असुरक्षित व्यवहारों को अवलोकन करने के लिए भी कहा गया, जिसका परिणाम निम्नानुसार है :

	पहली पारी		दूसरी पारी	
	सुरक्षित व्यवहार	असुरक्षित व्यवहार	सुरक्षित व्यवहार	असुरक्षित व्यवहार
पहला दिन	105	73	142	38
दूसरा दिन	64	33	82	15

यह परिणाम दर्शाता है कि बीबीएस प्रशिक्षण के कारण, पहली पारी से दूसरी पारी में सुरक्षित व्यवहारों की संख्या में बढ़ोतरी हुई है और असुरक्षित व्यवहारों में कमी आई है।

यह भी देखा गया कि प्रत्येक बीबीएस प्रशिक्षित पर्यवेक्षक द्वारा लगभग 1 प्रतिशत जोखिम-कमी में सहयोग दिया गया, जो दर्शाता है कि जितने अधिक पर्यवेक्षक होंगे, उतने ज्यादा अवलोकनों की संख्या होगी तथा संगठन में जोखिम-कमी का प्रतिशत उतना ही अधिक होगा।

तकनीशियनों एवं इंजीनियरों से पूछा गया कि उन्हें बीबीएस के विषय में सबसे अच्छा क्या लगा, जिसका परिणाम निम्नानुसार है: बीबीएस कामगारों के बीच आत्मनिरीक्षण तथा सतर्कता जगाता है। यह एक टीम-निर्माण तथा बढ़ते हुए परस्पर-संवाद की प्रक्रिया है। यह एक मानवीय संरक्षा है। बीबीएस में किसी को भी बदलने के लिए बाध्य नहीं किया जाता है। यह एक परस्पर संप्रेषण है। यह तक तकनीकी-आधारित दृष्टिकोण है। इसका तात्पर्य असुरक्षित व्यवहार को दंड देना ही नहीं है। हम व्यावहारिक परिवर्तन के द्वारा मनोवृत्ति बदल सकते हैं। यह सकारात्मक संरक्षा संस्कृति को बनाने में सहायक है। यह एक सुव्यवस्थित फीडबैक एप्रोच है।

तकनीशियनों एवं इंजीनियरों ने बताया कि पर्यवेक्षक अपने मासिक बैठकों में सुरक्षित व्यवहारों को प्रोत्साहित करने तथा सुरक्षित माहौल बनाने हेतु दोनों विषयों पर चर्चा करें। पर्यवेक्षकों के बीओएफपी के दौरान पर्यवेक्षितों की माइंड-रीडिंग नहीं करनी है। बीओएफपी को एक नियमित तरीके से करने की कोशिश करते रहे, ताकि कार्यस्थल में सुरक्षित तथा असुरक्षित व्यवहारों के प्रतिशत के अनुसार उनका संचयी प्रभाव सुनिश्चित किया जा सके। पर्यवेक्षकों को साप्ताहिक पर्यवेक्षण का कार्यक्रम पहले से ही बनाना होगा, ताकि वे अपने संगठन में विभागों के बीच बीओएफपी का विस्तार कर सके।

तकनीशियनों एवं इंजीनियरों ने यह जानने की भी इच्छा जाहिर की कि बीबीएस, विभिन्न प्रकारों के पीपीई को किस तरह नियमित किया जाए; क्या बीबीएस किसी के जीवन में भी उपयोगी है; पर्यवेक्षक और पर्यवेक्षित के बीच सहयोग कैसे प्राप्त करें; किसी पर्यवेक्षक के व्यक्तिगत तनाव को कैसे कम करें; क्या बीबीएस वास्तव में प्रभावी है; व्यवहार या मनोवृत्ति में से क्या सुधारें; बीबीएस के विमोचन हेतु कार्ययोजना कैसे तैयार करें; क्या बीबीएस को जोखिम भरे कार्य के वातावरण में कार्यान्वित किया जा सकता है; क्या बीबीएस ओएचएसएएस के अनुसार है; क्या प्रेक्षकों को समान ही रहना या परिवर्तित होना है।

बीबीएस चार्ट

बीबीएस संकल्पना	बीओएफपी: व्यवहारों का प्रेक्षण	बीओएफपी: फीडबैक संघटक
बीबीएस विशिष्ट व्यवहारों पर पर्यवेक्षण डाटा का संग्रहण है। बीबीएस ऊपर के स्तर से नहीं संचालित होता, बल्कि यह एक ऊर्ध्वगामी दृष्टिकोण है। बीबीएस का उद्देश्य है कि सुरक्षित एवं जोखिम-पूर्ण व्यवहारों की पहचान हो, चोटिल होने के अवसरों की पहचान हो, जोखिम के बारे में संप्रेषण हो, सुरक्षित समाधान की पहचान में सहायक हो।	पीपीई पीपीई का उपयोग करना उदा. चष्मे, श्रवण संरक्षा, दस्ताने, सख्त टोपी	विश्वास सकारात्मक अहंभाव सहनशक्ति विनोदशीलता सुनना मानवीय संबंध सक्रिय रहना मित्र न कि बॉस तनाव से आराम
बीबीएस की शुरूआत सभी कार्यक्षेत्रों तथा विभागों के लिए संक्षिप्त प्रशिक्षण सत्रों के द्वारा होती है। बीबीएस 'स्वयं की तथा दूसरों की संरक्षा' है।	**हाउसकीपिंग** कार्यक्षेत्र का उचित रखरखाव, कूड़ा-करकट तथा रद्दी माल का उठाया जाना, कोई बिखराव नहीं, रास्ते अबाधित नहीं।	**बीबीएस कार्यान्वयन** प्रबंधन से प्रतिबद्धता;
बीबीएस की मान्यता है कि प्रत्येक कर्मचारी संगठनात्मक संरक्षा में बदलाव ला सकता है।	**औजारों तथा उपकरण का प्रयोग** कार्य के लिए उचित औजारों का प्रयोग, औजारों का सही तरीके से प्रयोग तथा औजारों की अच्छी दशा में प्रयोग	इसे निम्न स्तर से परिचित कराना;
बीबीएस का मानना है कि जो परिमित है, वह संपूर्ण है।	**बॉडी पॉजिशनिंग/संरक्षी** शरीर के अंगों की पॉजिशनिंग/संरक्षा उदा. अग्नि से दूर रहना।	संचालन समिति का गठन कराना;

बीबीएस डाटा-संचालित संकल्प-निर्माण की एक प्रक्रिया है।	**सामग्री हैंडलिंग** उदा.लिफ्टिंग,पुशिंग एवं पुलिंग के दौरान बॉडी मैकेनिक्स	पर्यवेक्षकों का चयन तथा प्रशिक्षण, जांचसूची बनवाना;
बीबीएस का उद्देश्य संरक्षा नियमों को प्रवर्तित करना, परिवर्तन हेतु बाध्य करना, दूसरों के विषय में गप-शप करना, बॉस को रिपोर्ट करना नहीं है।	**संप्रेषण** मौखिक तथा अमौखिक पारस्परिक संवाद, जो कि संरक्षा को प्रभावित करते है	बीओएफपी के लिए क्षेत्रों/विभागों का चयन;
व्यवहार विषयपरक, परिभाष्य, पालनीय तथा परिमेय है।	**कार्यपद्धतियों का अनुसरण करना** उदा. परमिटों को प्राप्त करना, अनुपालन करना, एसओपी का अनुसरण करना, लॉक आउट, टैग-आउट कार्यपद्धतियां.........	समय सारिणी बनाना; बीबीएस पर्यवेक्षणों तथा बैठकों के लिए समय निर्धारित करना;
बीबीएस कंपनी-व्यापक, लोक-संचालित दृष्टिकोण है।	**दृश्य-केंद्रीकरण-सावधानी**	प्रलेखन, मूल्यांकन एवं प्रदर्शन; प्रगति के निर्धारण हेतु समीक्षा

कार्यक्षेत्र में बीबीएस कार्यान्वित करने हेतु दिशामार्ग

बीबीएस किसी संगठन में विभागों के बीच व्यक्तियों को बॉटम-अप अप्रोच के रूप में शामिल करना है। कार्यस्थल में बीबीएस कार्यान्वित करने तथा संरक्षा संस्कृति बनाने के उद्देश्यार्थ, हमें निम्नानुसार अनुसरण करना है:

1. सभी स्तरों पर प्रबंधक कार्मिकों के लिए बीबीएस पर एक जागरूकता कार्यक्रम।
2. संयंत्र पर कर्मचारियों का एक दिवसीय जागरूकता प्रशिक्षण।
3. बीबीएस प्रशिक्षित क्रमचारियों से (20 प्रतिशत) पर्यवेक्षकों का चयन करना तथा बीबीएस प्रशिक्षित कर्मचारियों में से 8-10 व्यक्तियों वाली संचालन समिति का गठन करना। संगठन के प्रति विभाग में से 8 पर्यवेक्षकों का चयन इस प्रकार किया जा सकता है कि एक पर्यवेक्षक द्वारा प्रत्येक व्यवहार की श्रेणी का पर्यवेक्षण किया जा सके।

4. पर्यवेक्षण-पद्धति को स्थापित करने, उपाय विकसित करने, परिशुद्धि तथा संगति की जांच करने, संचालन समिति का कार्य करने आदि हेतु संचालन समिति के सदस्यों तथा बीबीएस पर्यवेक्षकों का दो-दिवसीय कार्यस्थलीय व्यावहारिक प्रशिक्षण।
5. इस चरण पर बीबीएस पर्यवेक्षक तथा संचालन समिति के सदस्य कार्यस्थल पर बीबीएस लागू करने के लिए पूर्णतया तैयार हैं।
6. भारत में उपर्युक्त अभ्यास का संतोषजनक प्रयोग किया गया।

बीबीएस संचालन समिति के सदस्यों के लिए एक-दिवसीय कार्यशाला
संचालन समिति के सदस्यों के लिए इस कार्यशाला में निम्नलिखित घटक शामिल होंगे –
• प्रेक्षण की प्रक्रिया कैसे स्थापित करें।
• उपाय कैसे विकसित करें।
• परिशुद्धता एवं सुसंगति की जांच करना।
• बीबीएस समिति का कार्य करना।
• टीम की भूमिका।
• समस्या निवारण/संप्रेषण/फीडबैक प्रवीणता।
• मौजूदा संरक्षा स्थिति का एसडब्ल्यूओटी विश्लेषण।
• बौद्धिक पूर्ण प्रक्रिया।
• प्रस्तुतीकरण प्रवीणता

बीबीएस संचालन टीम का कार्य

बीबीएस संचालन टीम के सदस्य का मुख्य कार्य है कि वे प्रेक्षकों को उनके संबंधित संयंत्रों में नियमित रूप से परामर्श, मॉनीटर तथा प्रोत्साहित करें।

इस टीम के 15 कार्य उनकी नियमित संवीक्षा के लिए निम्नानुसार जांचसूची में सूचीबद्ध हैं :

जांचसूची बिंदु		हां	नहीं	टिप्पणियां
1	क्या पर्यवेक्षण की जांचसूची ठीक है(या संशोधन आवश्यक है)?			
2	क्या पर्यवेक्षक द्वारा दिन में एक बार अवलोकन पर्याप्त है ?			

3	क्या संयंत्र पर सुरक्षित/जोखिम-भरे व्यवहारों के प्रतिशत के संदर्भ में मासिक प्रगति है ?			
4	क्या बीबीएस संचालन टीम/पर्यवेक्षकों की पाक्षिक या मासिक बैठकें होती हैं ?			
5	क्या पर्यवेक्षक प्रतिदिन बीओएफपी करते हैं ?			
6	क्या पर्यवेक्षकों द्वारा डाटा की प्रविष्टि की जाती है ?			
7	क्या संयंत्र पर प्रति 10 दिनों में डाटा-विश्लेषण किया जाता है ?			
8	क्या संयंत्र पर बीबीएस की प्रगति चार्ट का प्रदर्शन किया जाता है ?			
9	क्या संयंत्र/क्यूएचएसई/मासं प्रमुखों द्वारा मासिक प्रगति रिपोर्ट प्राप्त की जाती है ?			
10	क्या संयंत्रों पर सर्वेक्षण रिपोर्ट कार्रवाइयां अवरूद्ध की जाती हैं ?			
11	क्या अलग-अलग पर्यवेक्षकों के मुद्दे लिखे जाते हैं ?			
12	क्या बाहरी समीक्षक द्वारा अनुवर्ती कार्रवाई मासिक आधार पर हो रही है ?			
13	क्या बीबीएस प्रोजेक्ट को बैनरों/पोस्टरों के द्वारा बीआईएलएजी में उचित प्रचार/विज्ञापन मिल रहा है ?			
14	क्या पर्यवेक्षकों को सम्मानित/महत्त्व दिया जाता है ?			
15	क्या उच्च प्रबंधन द्वारा सहयोग/समीक्षा/नेतृत्व हो रहा है ?			

संचालन टीम समीक्षा/चर्चा हेतु प्रतिमाह में एक या दो बार बैठक लेगी।

बीबीएस परियोजना संगठन : एक उदाहरण

बीएफएल पुणे 'सुरक्षा मार्ग'

क्र.सं.	गतिविधि	उत्तरदायित्व	
	विमोचन होने से पूर्व की गतिविधियां		
1.	बीएफएल हेतु बीबीएस शीर्ष-टीम का गठन।	टीम-ए	टीम-बी
2.	पर्यवेक्षकों तथा एम एवं ई अन्य हेतु	पीएमटी/	एमएसआर/

	प्रशिक्षण मॉड्यूल तैयार करना।	डीएलडी/पीबीपी	एसएमएस
3.	पर्यवेक्षण जांचसूची की डिजाइन एवं कस्टमाइजेशन का कार्य	पीएमटी/ डीएलडी/पीबीपी	एमएसआर/ एसएमएस
4.	सभी कर्मचारियों तथा ठेका श्रमिकों को जागरूकता प्रशिक्षण (प्रशिक्षण अवधि - 02 घंटे)	सभी सदस्य	
5.	पर्यवेक्षकों का चयन	सभी सदस्य	
6.	पर्यवेक्षकों के लिए प्रशिक्षण कार्यक्रम	सभी सदस्य	
7.	उप समितियों का गठन तथा उद्घोषणा-विभाग वार	सभी सदस्य	
8.	टीम वार साप्ताहिक बैठक	एसीजे	एसडीपी
9.	शीर्ष-टीम सदस्य सभा-पाक्षिक	एसीजे	एसडीपी
10.	बीबीएस कार्यक्रम का प्रचार-प्रसार – टीबीटी चेतावनी योजना(एलपी)/पोस्टर्स/व्यवहार श्रेणियां स्टीकर्स की तैयारी	एसबीएस	एसडीपी
	कार्यान्वयन		
1.	प्रारंभ समारोह	पीबीपी	एमएसआर
2.	पर्यवेक्षकों के डाटा विश्लेषण	बीपीबी/ आरपीए	डीएनजी/ एपीके/एमएनबी
3.	जोखिम-भरे व्यवहारों का वर्गीकरण(परेटो विश्लेषण)	डीएलडी/एसीजे	
4.	एचआईआरए के साथ असुरक्षित शर्तों तथा संयोजन का प्रेक्षण	सभी सदस्य	
5.	उपयुक्त नियंत्रण उपायों के लिए कार्य योजना	सभी सदस्य(एचएसई लीडरों को शामिल करते हुए)	
6.	मासिक विभागीय बैठकों के द्वारा जोखिम-भरे व्यवहारों की समीक्षा	विभाग-संयोजक	
7.	बारंबार काउंसिलिंग के द्वारा जोखिम	सभी पर्यवेक्षक	

	भरे व्यवहारों में सुधार		
8.	उच्च % सुरक्षित व्यवहार दर्शाने वाले विभागों/पर्यवेक्षकों को मान्य करना	शीर्ष-टीम	
9.	संरक्षा सूचनापट्ट पर सभी विभागों के सुरक्षित/जोखिम-भरे व्यवहारों के ग्राफ का प्रदर्शन	विभागीय संयोजक	
10.	सुरक्षित व्यवहारों का न्यूनतम % तथा कार्य संरक्षा पर्यवेक्षण से सुयोजित तथा संरक्षा प्रतिदर्शी तकनीक वाले विभागों/अनुभागों को महत्व देना/केंद्रित करना	शीर्ष टीम के सहयोग से पर्यवेक्षक	
11.	मासिक डाटा विश्लेषण(विभाग वार तथा बीएफएल स्तर)	शीर्ष टीम	
12.	बीबीएस आंतरिक लेखा परीक्षा – टीम के बीच	टीम –बी	टीम-ए
13.	आंतरिक लेखापरीक्षा जॉच के अनुसार गैर-अनुपालन की स्वीकृति	टीम-ए	टीम –बी
14.	बीबीएस आवर्तन पुनरावृत्ति – सुरक्षित व्यवहार- 85% तक (आधार स्तर – 60%)		
15.	एचओडी एवं लाइन मैनेजरों(50) के लिए बीबीएस प्रशिक्षण	टीम-ए एवं बी लीडर उपलब्ध रहें	
16.	कामगारों के लिए बीबीएस प्रशिक्षण(शीर्ष टीम हेतु समर्थन समूह)		
17.	शीर्ष टीम के मार्गदर्शन में समर्थन समूह द्वारा बीबीएस व्यावहारिक सर्वेक्षण	एसवीपी/ वीबीपी	आरडीबी/ एसीजे
18.	शीर्ष टीम द्वारा सर्वेक्षण प्रस्तुति	टीम-ए एवं बी लीडर	
19.	शीर्ष टीम द्वारा बीएफएल में सर्वेक्षण के		

	जरिए पाक्षिक भ्रमण		
20.	बीबीएस प्रगति के बारे में एचओडी तथा ईडी को प्रतिमाह रिपोर्टिंग		
	विभागों की संख्या – 16		
	प्रशिक्षकों की संख्या – 20		
	कर्मचारियों की संख्या – 6800		
	प्रति प्रशिक्षक कर्मचारी – 340		
	टीम लीडर- ए – प्रशांत पांचाल		
	टीम लीडर – बी – मोहन्न राजोबा		
	सभी शीर्ष बीबीएस टीम को परेटो विश्लेषण प्रशिक्षण दिनेश दौंडकर द्वारा(25/02/10 को अपराह्न 0500 बजे)		
	शीर्ष टीम द्वारा लोगो एवं मोटो तय किया जाए	एसवीएस/आरपीए	
	संयोजक -		
	टीम – ए – पीएमटी		
	टीम – बी – एसएमएस		
	बीबीएस सलाहकार : केकेके, एचएलके		

बीबीएस कार्यान्वयन में आने वाली बाधा :

एक बड़ी इंजीनियरिंग कंपनी के संरक्षा सलाहकार - 'उत्पादन में थोड़ी देर है, इसीलिए हमने बीबीएस को स्थगित किया है। एक बार उत्पादन की तिमाही समाप्त हो जाए, हम बीबीएस की पुनः समीक्षा करेंगे।'

ईएसएम कनाडा लिमिटेड – एक शिपिंग कंपनी के लिए बीबीएस कार्यान्वयन योजना(अप्रैल 2010 से मार्च 2011)

ए) पूर्व कार्यान्वयन गतिविधियां/सूची	
1.	बीबीएस पर दिनांक 09 फरवरी 10 को मुख्यालय में चर्चा।
2.	दिनांक 26/27 फरवरी 10 को मुंबई में समन्वयक हेतु दो दिवसीय बीबीएस प्रशिक्षण कार्यक्रम।
3.	प्रबंधक समिति के साथ परामर्श के पश्चात बीबीएस पर्यवेक्षण जांचसूची प्रचलित

	करना।(नमूना संलग्न)
4.	पर्यवेक्षण डाटा एकत्रीकरण, डाटा विश्लेषण तथा कंपनी कार्यालयों तथा जहाज़ों में प्रदर्शन हेतु बीबीएस वर्गीकरण।
5.	दिनांक 31 मार्च से 02 अप्रैल 2010 को शंघाई में कंपनी सेमिनार के दौरान कंपनी निदेशकों, प्रबंधकों तथा जहाज़ मंडल के वरिष्ठ अधिकारियों को बीबीएस संकल्पना का परिचय।
6.	शंघाई में दिनांक 03 अप्रैल 10 को अधीक्षकों तथा वरिष्ठ अधिकारियों को बीओएफपी(व्यवहार प्रेक्षण फीडबैक प्रक्रिया) पर बीबीएस वर्कशॉप। अधीक्षकों को बीबीएस प्रशिक्षक तथा प्रेक्षक के रूप में कार्य करना होगा।
7.	03/04/10 को शंघाई में प्रबंधन समिति से बीबीएस कार्यान्वयन योजना पर चर्चा तथा अनुमोदन।(कंपनी के आईएसएम/एसईपी प्रणाली में बीबीएस एकीकृत करना। प्रति पर्यवेक्षक प्रति सप्ताह अपेक्षित पर्यवेक्षणों की संख्या-2/बीबीएस पर्यवेक्षकों के लिए प्रोत्साहन योजना पर विचार–एक भत्ता, अधिकतम अवलोकन करने वाले जहाज़ों को पुरस्कृत करना/उच्चतम% सुरक्षित व्यवहार।)
8.	परियोजना का नाम तय करना(उदा. बीबीएस कार्यान्वयन/बीबीएस-नाविक हेतु सुरक्षित मार्ग)।
9.	एम.डी. द्वारा सेमिनार निष्पादन दिवस पर बीबीएस लांच की औपचारिक उद्घोषणा।
बी) बोर्ड पर कार्यान्वयन	
1.	अप्रैल के द्वितीय सप्ताह से संबंधित प्रशिक्षण द्वारा बोर्ड पर सभी क्रू को बीबीएस जागरूकता प्रशिक्षण(अवधि – 2 घंटे)।
2.	उपर्युक्तानुसार, संबंधित प्रशिक्षक द्वारा बोर्ड पर बीबीएस पर्यवेक्षक प्रशिक्षण (अर्धदिवसीय)।
3.	6 महीने में बीबीएस जागरूकता तथा पर्यवेक्षक प्रशिक्षण पूर्ण करना(सितंबर 10 के अंत तक)।
4.	संयोजक द्वारा प्रशिक्षण सामग्री(पावरपाइंट प्रस्तुति व पठन सामग्री)तैयार करना तथा वितरित करना।
5.	जहाजों के लिए प्रदर्शन बैनर तैयार करना(बीबीएस – सुरक्षित कार्य वातावरण के प्रति आश्वासन/)
6.	शिपबोर्ड एसईपी समिति को बीबीएस संचालन समिति के रूप में अंगीकृत करना। पर्यवेक्षकों की न्यूनतम संख्या 5-6 हो(उदा. टॉप 4, कनिष्ठ अधिकारी, बोसन)। एसईपी समिति की बैठक के कार्यवृत्त में बीबीएस गतिविधियां शामिल

	करना।
7.	प्रति प्रेक्षक प्रति सप्ताह न्यूनतम अपेक्षित पर्यवेक्षण(2)।
8.	सप्ताह के अंत में अवलोकन डाटा का संयोजक को हस्तांतरण करना।
9.	संयोजक द्वारा बीबीएस डाटा का विश्लेषण करना तथा मासिक रिपोर्ट तैयार करना – जिसे कि जहाजों तथा कंपनी को वितरित करना। बोर्ड तथा कार्यालयों में सुरक्षित/जोखिम भरे व्यवहारों का ग्राफिकल प्रदर्शन करना।
10.	संयोजक द्वारा रिपोर्ट किए गए सुरक्षित तथा जोखिम-भरे व्यवहार तथा असुरक्षित स्थितियों की प्रवृत्ति मॉनीटर करना/संयोजक द्वारा उपयुक्त समयोजित कार्रवाइयों के लिए संरक्षा विभाग तथा वैसल के साथ समन्वय करना।
11.	बोर्ड पर एसईपी समिति की बैठक के दौरान बीबीएस रिपोर्ट पर चर्चा।
12.	नौसेना में बीबीएस लागू करने तथा समीक्षा करने हेतु जहाजों पर प्रशिक्षकों द्वारा अनुवर्ती दौरा(अक्टू 10 से मार्च 11)

क्षेत्र प्रेक्षणों की रिपोर्टें :

बड़ी इंजीनियरिंग मल्टी-नेशनल संगठन का मामला :

जनवरी 2010 के दौरान, पुणे में 5000 से अधिक कार्यरत कर्मचारियों वाली एक बड़ी इंजीनियरिंग संगठन के 22 संरक्षा सह-संयोजकों का एक समूह बीबीएस के तीन दिवसीय प्रशिक्षण के लिए गया। उन्होंने रिपोर्ट की कि वहां प्रति कर्मचारी औसतन 4 असुरक्षित व्यवहार होते हैं तथा निर्णय लिया कि प्रत्येक संरक्षा संयोजक एक-एक विभाग को अंगीकृत करेंगे तथा व्यक्तियों के बीच बीबीएस जागरूकता पैदा करेंगे। उन्होंने प्रत्येक पर्यवेक्षक के उत्तरदायित्व तथा भूमिका पर विशेष बल दिया, जो कि बीबीएस परियोजना का मूल ढांचा है। नवीन प्रशिक्षित पर्यवेक्षक असुरक्षित व्यवहारों के अवलोकन में ही अभ्यस्त थे तथा पुरानी मानसिकता के चलते फीडबैक नहीं देते थे। यह महसूस किया गया कि पर्यवेक्षकों को शोरगुल वाले कार्य क्षेत्र में फीडबैंक नहीं देना चाहिए, बल्कि पर्यवेक्षकों को शोरगुल वाले क्षेत्र से बाहर ले जाएं, ताकि संप्रेषण पूर्णतया सुना जा सके। हालांकि पर्यवेक्षक का काम पर्यवेक्षण करना है तथा एक समय पर एक श्रमिक से फीडबैक लेना है, परंतु उसे ऐसे अन्य लोगों पर भी अवलोकन करना होगा, जिन्होंने पीपीई नहीं पहना है, पर्यवेक्षण दौरे के दौरान दूसरों का भी अवलोकन करना उसका कार्य है। पर्यवेक्षक को बीओएफपी के दौरान पर्यवेक्षण किए जाने वाले

व्यक्ति का नाम पूछने की भी आवश्यकता नहीं है। पर्यवेक्षकों को अवलोकन किए जाने वाले व्यक्तियों पर सुरक्षित व्यवहारों को बनाए रखना तथा उन्हें इस बाबत समझाना चाहिए।

पर्यवेक्षक की ड्यूटी पर्यवेक्षण दौरे के दौरान उसके द्वारा अवलोकित/अनुवर्ती असुरक्षित परिस्थितियों को बंद करना भी है। यदि ये असुरक्षित परिस्थितियां किसी पर्यवेक्षक के सीमा क्षेत्र से परे हैं, तब उसे इंजीनियरिंग/अनुरक्षण विभाग से संपर्क करना चाहिए। पर्यवेक्षकों की बैठक के दौरान प्रत्येक पर्यवेक्षक को आंतरिक विचार-विमर्श हेतु असुरक्षित परिस्थितियों की फोटो खींचने के लिए डिजिटल कैमरा लाना चाहिए।

बीबीएस पर्यवेक्षकों के द्वारा रिपोर्ट किए गए बीबीएस व्यावहारिक प्रशिक्षण के लाभ :

1. पर्यवेक्षकों के कोचिंग तथा फीडबैक पटुता में सुधार।
2. पर्यवेक्षण किए जाने वाले व्यक्तियों की ओर से एसओपी से विसामान्यता में सुधार।
3. असुरक्षित परिस्थितियों के विषय में अवलोकन किए जाने व्यक्तियों की जानकारी में बढ़ोतरी।
4. संरक्षा वृद्धि के विषय में पर्यवेक्षक तथा प्रेक्षितों की जानकारी।
5. पर्यवेक्षकों की मानसिकता में सकारात्मक परिवर्तन।
6. पर्यवेक्षकों द्वारा दिनचर्या में बीओएफपी बनाए रखने के कारण पर्यवेक्षण में लगने वाले समय में कमी।
7. प्रेक्षित ज्यादा जागरूक हो गए।
8. पर्यवेक्षक के आत्मविश्वास तथा जागरूकता में वृद्धि।
9. प्रेक्षितों के आस-पास प्रचालक जागरूक हो गए।
10. प्रेक्षकों तथा प्रेक्षितों के बीच रिश्तों में सुधार
11. प्रेक्षकों की आत्म-संतुष्टि में बढ़ोतरी। "हम कुछ अलग कर रहे हैं।"
12. बीओएफपी में प्रेक्षितों के विश्वास में वृद्धि।

पर्यवेक्षकों ने चर्चा की कि किस प्रकार जोखिम-भरे/असुरक्षित व्यवहारों को क्षति के वित्तीय डाटा से जोड़ा जा सकता है? ऐसा बीबीएस लाने से पहले वाले वर्षों में खर्च हुई क्षति की लागत, गंवाएँ गए श्रम-दिवस, संपत्ति क्षति, कोर्ट के मामले, क्षतिपूर्ति, चिकित्सा व्यय की गणना के द्वारा किया जा सकता है, जिसकि तुलना बीबीएस लाने के बाद के वर्षों में बचाई गई ऐसी लागतों से की जा सकती है।

मल्टी-नेशनल केमिकल की एक इकाई का मामला : असुरक्षित व्यवहार

फरवरी 2010 के दौरान, हैदराबाद में एक मल्टी-नेशनल केमिकल यूनिट के माध्यम से प्रबंधन स्तर के 20 प्रतिभागियों के एक समूह ने बीबीएस कार्यशाला में भाग लिया और उन्होंने पाया कि प्लांट में कार्यरत प्रचालकों के असुरक्षित व्यवहारों की संख्या 71 है तथा प्रति कर्मचारी औसतन 3.5 असुरिक्षत व्यवहार(71/20 =3.5) पाए गए, जिसका तात्पर्य है कि वर्तमान में वहां असुरक्षित व्यवहारों (62 कर्मचारी X 3.5) की कुल संख्या 217 है। बीबीएस प्रशिक्षण कार्यक्रम के दौरान एक प्लांट मैनेजर ने बताया कि – "हम वर्ष 2008 से बीबीएस शुरू करने के लिए इच्छुक नहीं थे। हमारी सोच थी कि यह एक और कार्य बढा देगा, तथापि हमने बेमन इसकी शुरूआत की। प्रशिक्षण तथा इस संबंध में साहित्य पढ़ने के बाद हमें विश्वास हुआ कि हम यह कर सकते हैं।" कंपनी के उत्पादन प्रमुख ने बताया-"बीबीएस बिना किसी अतिरिक्त प्रयासों के 100 प्रतिशत संरक्षा दे सकता है।"

एक इंडियन केमिकल ग्रुप ऑफ कंपनीज का मामला

अप्रैल 2010 में, इंडियन केमिकल ग्रुप ऑफ कंपनीज के 41 वरिष्ठ प्रबंधकों ने मुंबई में एक पूर्ण-दिवसीय बीबीएस कार्यशाला में भाग लिया। कंपनियों के एक अधिशासी निदेशक ने अपने शुरूआती संबोधन में कहा- "प्रत्येक व्यक्ति संरक्षा का एक केंद्र है।" उन्होंने यह भी कहा कि यदि हम संरक्षा की स्थिति में बदलाव चाहते हैं, तो सबसे पहले हमारे प्रबंधकों को शिक्षित करना होगा। प्रबंधकों ने राय दी कि संगठनों में संरक्षा के परिदृश्य में सुधार करने हेतु कुछ भय या दंड का प्रावधान किया जाना चाहिए। इन प्रबंधकों द्वारा पिछले एक वर्ष के दौरान प्रति कर्मचारी असुरक्षित व्यवहारों की औसतन संख्या 05 पाई गई (कुल असुरक्षित व्यवहार 197/41 = 5), जिससे उनके संगठनों में बीबीएस की आवश्यकता महसूस की गई। एक प्रबंधक को पता चला कि बीबीएस को कार्यनिष्पादन मूल्यांकन के साथ जोड़ने की आवश्यकता है, क्योंकि पर्यवेक्षक अपनी पहल तथा स्वयंसेवा के जरिए संगठन के संरक्षा सुधार के प्रति योगदान देते हैं।

बीबीएस कार्यक्रम में प्रतिभागियों की प्रतिक्रिया

विश्वनाथ दिवेकर – प्रत्येक का अभिप्राय सकारात्मक है। (+Ve मनोविज्ञान)

कृष्णराज – मानवीय भूल-सुधार हेतु अच्छा साधन।

मकरंद – व्यक्तिगत व्यवहार हेतु नैतिक महत्ता।

विशाल अर्जेल – इसे कार्यात्मक रूप में देखना है।

हेमंत काले – समारोह में प्रतिभागिता।

सुब्रमणियन	– हमें पता है हम कर सकते हैं।
मुथू	– देख-भाल के साथ-साथ।
मारूति गांवकर	– प्रोत्साहन।
अर्धेंदु	– अपने आपको बदलना।
सुजीत घोष	– निवारक प्रक्रिया(निष्कासन की जड़ तक)।
चक्रपाणी डी	– संरक्षा संस्कृति के प्रति यथार्थवादी पहुंच।
कुंडू	– संस्कृति बदलने में प्रचालकों की मानसिकता बदलने में प्रयोगी।
नीतिन उमालकर	– इसने हमें रोगी से डॉक्टर बना दिया।
वाघ ए.वी.	– संरक्षा संस्कृति बनाने में सरल, योग्य तथा आसान।
थॉमस एन	– लांच करने का सही समय तथा मैं परिवर्तन ला सकता हूँ।
लीओ	– सर्वोत्तम तथा सरल एवं प्रभावशील।

<u>एक मल्टी-नेशनल फार्म इक्विपमेंट सेक्टर का मामला</u>

जुलाई 2010 के महीने के दौरान महाराष्ट्र राज्य में विभिन्न विभागों से 81 कार्यकारियों ने चार एक-दिवसीय बीबीएस कार्यशालाओं में भाग लिया। इसमें प्लांट के प्रमुख ने बताया कि उनके यहां पिछले 7 वर्षों में कोई दुर्घटना नहीं हुई थी, किंतु बहुत सारी क्षति हुई थी। ओएचएसएएस लेखापरीक्षा के दौरान, इस पर विशेष बल दिया गया कि सभी अधिकारियों तथा श्रमिकों को बीबीएस से परिचित कराया जाया। इसीलिए उपाध्यक्ष ने संगठन में बीबीएस लांच करने का निर्णय लिया।

पिछले एक वर्ष के दौरान कार्यकारियों द्वारा पर्यवेक्षण किए गए असुरक्षित व्यवहारों की संख्या 09 थी, जिसे कि कर्मचारियों की कुल संख्या से गुणा किया गया अर्थात 2000 X 9= 18,000/-

प्रति असुरक्षित व्यवहार की लागत की गणना निम्नलिखित फॉमूर्ला के अनुसार की जाती है :

पिछले वर्ष की इंजुरी की संख्या × प्रति इंजुरी अनंतिम लागत/

असुरक्षित व्यवहारों की संख्या

= 55 × 10,000 +(रु. 2 लाख की लागतवाली एक बड़ी इंजुरी)/18,000 = रु. 42 प्रति असुरक्षित व्यवहार

इसका मतलब कि यदि संयंत्र में 18,000 असुरक्षित व्यवहार होते हैं, तब वहां बीबीएस की आवश्यकता है तथा संगठन में बीबीएस की शुरूआत करने से न केवल कर्मचारियों को होने वाली इंजुरी की संख्या में आती है, बल्कि 18,000 असुरक्षित व्यवहारों की मौजूदगी के कारण प्रति दिन रु. 7.5 लाख की लागत भी बच जाती है।

वर्कशॉप में आए प्रतिभागियों ने निम्नलिखित मुद्दों पर चर्चा की :

ए) चार प्रकार के प्रेक्षित : अनिच्छुक नौसिखिये, उच्च जोखिम लेने वाले, रफ-टफ श्रमिक(सामान्यतया ये तीनों प्रकार 7% वाले होते हैं) एवं चौथे प्रकार के वे सहयोगी श्रमिक होते हैं, जिनका प्रतिशत 93% है। सुरक्षित व्यवहार बनाए रखने के लिए 7% वाली श्रेणी को नियमित पर्यवेक्षण में रखा जाना चाहिए।

बी) केवल कर्मचारियों के बीच जागरूकता पैदा करने से ही संपूर्ण संरक्षा सुनिश्चित नहीं की जा सकती है। असुरक्षित व्यवहारों में सुधार आवश्यक है। हालांकि शून्य असुरक्षित व्यवहार बीबीएस का केंद्र है, परंतु नियमित संरक्षा शिक्षा तथा प्रशिक्षण के द्वारा कर्मचारियों के बीच संरक्षा अभिवृत्ति, मान्यता तथा जानकारी पैदा करना भी उतना ही महत्वपूर्ण है। दुर्घटना/घटना के उद्देश्यों को शून्य करने के अलावा संरक्षा संस्कृति का निर्माण भी बीबीएस के उद्देश्य में शामिल हैं।

सी) बीबीएस कार्यान्वयन के चार चरण : सभी विभागों के प्रमुखों को एक-दिवसीय बीबीएस जागरूकता प्रशिक्षण, प्रत्येक विभाग से एक प्रशिक्षक को दो-दिवसीय प्रशिक्षण, ताकि वे बदले के सभी कर्मचारियों को जागरूकता प्रशिक्षण दे सकें, बीबीएस के ऐसे पर्यवेक्षकों को एक-दिवसीय प्रशिक्षण जो कि कर्मचारियों के जागरूकता प्रशिक्षण समूह से बने हैं, प्रशिक्षित पर्यवेक्षकों से बनी बीबीएस संचालन समिति को एक-दिवसीय प्रशिक्षण।1

<u>मल्टी-नेशनल ऑटोमोबाइल कंपनी का मामला</u>

अगस्त 2010 के महीने के दौरान महाराष्ट्र राज्य में एक ऑटोमोबाइल कंपनी के 10 विभागों से 23 प्रबंधकों की एक टीम ने एक-दिवसीय कार्यशाला में भाग लिया।

पिछले एक वर्ष के दौरान कार्यकारियों द्वारा पर्यवेक्षण किए गए असुरक्षित व्यवहारों की संख्या 06 थी, जिसे कि कर्मचारियों की कुल संख्या से गुणा किया गया

अर्थात 1700 × 6 = 10,200/-

संगठन में पिछले वर्ष के दौरान होने वाली इंजुरी का विवरण:

72 प्रथमोपचार चोट × लागत अर्थात रु. 10,000/- प्रति चोट = रु. 7,20,000

34 अप्रतिवेद्य चोट × लागत अर्थात रु. 5,000/- प्रति चोट = रु.1,70,000

6 प्रतिवेद्य चोट × लागत अर्थात रु. 10,000/- प्रति चोट = रु.60,000

चोटों की कुल संख्या = 112, चोटों की कुल लागत = रु. 9,50,000/-

प्रति चोट लागत = रु. 8,482

प्रति असुरक्षित व्यवहार की लागत की गणना निम्नलिखित फॉमूर्ला के अनुसार की जाती है :

पिछ्ले वर्ष की इंजुरी की संख्या X प्रति इंजुरी अनंतिम लागत/

असुरक्षित व्यवहारों की संख्या

= 112 × रु. 8,482 / 10,200 = रु. 93 प्रति असुरक्षित व्यवहार

असुरक्षित व्यवहारों की लागत बड़ी है, जो कि पिछले वर्ष चोटों की संख्या, प्रति चोट अनंतिम लागत, तथा पर्यवेक्षण किए गए असुरक्षित व्यवहारों की संख्या के आधार पर संगठन से संगठन में बदलता है।

बीबीएस प्रशिक्षित पर्यवेक्षक किसी प्रेक्षण भ्रमण के दौरान प्रेक्षितों पर पर्यवेक्षण के द्वारा असुरक्षित व्यवहार को जहां-तहां बंद कर देते हैं।

ऑइल कॉर्पोरेशन का एक मामला :

मुंबई में जुलाई-अगस्त 2010 के दौरान भारत से विभिन्न मार्केटिंग विभागों के 36 प्रमुखों के व्यवहार संरक्षा पर दो दिवसीय वर्कशॉप में हिस्सा लिया। उनके कार्यक्षेत्र में निम्नलिखित असुरक्षित/जोखिम-भरे व्यवहार पाए गए :

ए. 30% कर्मचारियों द्वारा संरक्षा जूतों का प्रयोग नहीं किया जाता।

बी. 100 % कर्मचारियों द्वारा हेलमेट का प्रयोग नहीं किया जाता।

सी. 100 % कर्मचारियों द्वारा दस्तानों का प्रयोग नहीं किया जाता।

डी. 70% कर्मचारियों द्वारा मोबाइल का प्रयोग नहीं किया जाता।

ई. 70% कर्मचारियों द्वारा संरक्षा बेल्टों का प्रयोग नहीं किया जाता।

एफ. 30% कर्मचारियों द्वारा संरक्षा समिति की बैठकों में हिस्सा नहीं लिया जाता।

जी. 5% कर्मचारियों द्वारा प्रतिबंधित क्षेत्र में धूम्रपान।

एच. 100% कर्मचारियों द्वारा गॉगल्स(चष्मों) का प्रयोग नहीं किया जाता।

उन्होंने बताया कि अच्छे संरक्षा प्रबंधन के लिए व्यवहार संरक्षा एक नयी संकल्पना तथा पद्धति है, जो कि एक सार्थक प्रयास है, मूल आधार पर कार्यान्वयन के योग्य है तथा जो व्यावहारिक है।

एक इंजीनियरिंग प्रापण निर्माण कंपनी का मामला

अगस्त 2010 में 2000 कर्मचारियों से नियुक्त एक एमएनसी कंपनी के 18 वरिष्ठ प्रबंधन कार्मिकों ने एक पूर्ण दिवसीय व्यवहार संरक्षा कार्यशाला में भाग लिया। उन्होंने

पाया कि उनके कर्मचारियों के बीच औसतन 5 असुरक्षित व्यवहार होते हैं। इस तरह सभी कर्मचारियों के लिए असुरक्षित व्यवहारों की कुल संख्या 10,000 होगी।

उन्होंने पाया कि कार्यालय कि संरक्षा दृष्टिकोण से असुरक्षित व्यवहारों की छह श्रेणियां हो सकती हैं : हाउसकीपिंग(जल छलकाव, फाइलें अस्त-व्यस्त) उपकरणों का प्रयोग(कंप्यूटर/लैपटॉप), शरीर की अवस्था(गलत अवस्था), संप्रेषण(जोर-जोर से बोलना, दूसरों के बीच हस्तक्षेप करना), दृष्टि एकाग्र करना तथा मोबाइल का प्रयोग करना।

उन्होंने निम्नानुसार प्रति असुरक्षित व्यवहार की गणना रु. 50 की है :

पहली 50 प्रथमोपचार क्षति @ रु. 1000/- प्रति क्षति = कुल रु. 50,000

3 लॉस्ट टाइम इंजुरी @ रु. 20,000/- प्रति इंजुरी = कुल रु. 60,000

7 बाल-बाल बचे इंजुरी @ रु. 5000/- प्रति = कुल रु. 35,000

कुल लागत :

= रु. 1,45,000

प्रतिभागियों ने पाया कि चूंकि ये सभी इंजुरी तथा बाल-बाल बचे दुर्घटनाएं पिछले वर्ष परियोजना के पूर्ण होने के समय की थी, अतः इनसे वास्तविक स्थिति साफ नहीं हो पाएगी। परियोजना के दौरान, कुल लागत तकरीबन रु. 5 लाख होगा। अतः प्रति असुरक्षित व्यवहार की लागत की गणना रु. 50/- है। (रु. 5 लाख/10,000 असुरक्षित व्यवहार)

उन्होंने एक कार्ययोजना बनायी : संचालन समिति(10 सदस्य), क्षेत्र बीबीएस समन्वयक(20 सदस्य) तथा बीबीएस प्रेक्षक(200 सदस्य)।

प्रशिक्षक को प्रशिक्षण कार्यक्रम : संगठन के भीतर ही बीबीएस परामर्शदाताओं का निर्माण :

वे सभी जिन्होंने बीबीएस पर एक-दिवसीय जागरूकता कार्यशाला में भाग लिया था, वे दो दिनों के लिए 'प्रशिक्षक को प्रशिक्षण' कार्यक्रम में भाग लेने के पात्र हैं, जिसमें निम्नलिखित शामिल हैं :-

पहला दिन : पूर्वा. 9.30 से अप. 4.30 बजे तक

1. व्यवहार आधारित प्रशिक्षक : एक परिचय
2. व्यवहार आधारित प्रशिक्षक हेतु अपेक्षित निपुणता
3. बीबीएस पर विषयगत जानकारी परीक्षा

4. बीबीएस पर प्रयोग

दूसरा दिन : पूर्वाह्न 9.30 बजे से अपराह्न 4.30 बजे तक

1. प्रशिक्षकों के लिए मानक मैनुअल विकसित करना
2. बीबीएस प्रशिक्षकों की पहचान/एकीकरण
3. विभाग-वार प्रत्येक प्रशिक्षक हेतु कार्य योजना
4. विदाई कार्यक्रम।

बीबीएस कार्यशाला के प्रतिभागियों के कुछ विचार

ए.	व्यवहार संरक्षा असुरक्षित व्यवहारों को देखने का एक गॉगल(चष्मा) है।
बी.	हम मानव-संबंधित असुरक्षित व्यवहारों को अवलोकन हेतु जेएसए में बीबीएस दृष्टिकोण को शामिल करेंगे।
सी.	जब किसी संगठन में सभी प्रयासों से संरक्षा की स्थिति नहीं सुधरती, तब व्यवहार संरक्षा मददगार है।
डी.	बीबीएस के विषय में प्रबंधकों की पुरानी मान्यता के अनुसार निम्नलिखित प्रश्न उठ सकते हैं : क्या हम कुछ हद तक टॉप-डाउन अप्रोच का उपयोग कर सकते हैं? यदि बीबीएस कोचिंग काम नहीं कर रही है, तो क्या हम किसी तकनीक को बलपूर्वक/दंडपूर्वक लागू कर सकते हैं?
ई.	व्यवहार संरक्षा पीयर-आधारित है, न कि भय-आधारित।
बीबीएस भारतीय अनुसंधान से सीखे गए अध्याय निम्नानुसार हैं :-	
ए.	असुरक्षित व्यवहारों की श्रृं,खला 'बाल-बाल बचे' दुर्घटना, इंजुरी, फैटेलिटी से अधिक है।
बी.	किसी भी कार्यस्थल में संगठन के आकार के आधार पर दैनिक असुरक्षित व्यवहारों की संख्या सैकड़ों/हजारों होती है।
सी.	शून्य दुर्घटना रेकॉर्ड तथा अंतर्राष्ट्रीय प्रमाणन प्राप्त करने से वास्तव में संगठन की सुरक्षा सुनिश्चित नहीं की जा सकती, जब तक कि हमारा लक्ष्य कार्यस्थलों में असुरक्षित व्यवहारों को शून्य करना नहीं होता है।
डी.	बीबीएस हस्तक्षेप से असुरक्षित व्यवहारों में कमी तथा सुरक्षित व्यवहारों में बढोतरी देखी गई है। बीबीएस प्रशिक्षण से संगठन में असुरक्षित परिस्थितियों की संख्या कम करने में भी मदद मिलती है।
ई.	प्रबंधन वर्ग ने यह मानना शुरू कर दिया है कि केवल इंजीनियरिंग नियंत्रण से

	ही कार्यस्थल में पर्याप्त सुरक्षा मिलती है, जब तक कि कार्यस्थलों में संपूर्ण सुरक्षा सुनिश्चित करने की दिशा में असुरक्षित व्यवहारों पर नियंत्रण नहीं किया जाता तथा व्यवहार-संरक्षा का अभ्यास नहीं किया जाता है।
एफ.	हालांकि ओएचएसएएस 18001:2007 में संरक्षा के व्यवहार-वादी पहलुओं पर जोर देने वाले तीन शर्तें शामिल की गई हैं, संगठनों ने अब तक इस पर अनुपालन नहीं किया है, चूंकि ओएचएसएएस 18001:2007 में इन शर्तों को किस प्रकार कार्यान्वित किया जाए, इस संबंध में कोई दिशानिर्देश नहीं दिए गए हैं।
जी.	पिछले वर्षों की तुलना में भारतीय बहुराष्ट्रीय संगठनों में कार्यस्थल पर संरक्षा मे मानवीय व्यवहार के पहलुओं पर और अधिक ध्यान देना शुरू किया है।
एच.	कार्यस्थल पर बीबीएस का कार्यान्वयन करने तथा संरक्षा संस्कृति बनाने की दिशा में, निम्नलिखित कदम उठाए जाने चाहिए- 1. सभी स्तरों पर प्रबंधन कार्मिकों के लिए बीबीएस पर एक जागरूकता कार्यक्रम का आयोजन करें। 2. संयंत्र में कर्मचारियों के लिए जागरूकता प्रशिक्षण का आयोजन करें। 3. बीबीएस प्रशिक्षित कर्मचारियों में से संचालन समिति का गठन करें तथा पर्यवेक्षकों को चयन करें। 4. पर्यवेक्षण प्रक्रिया किस प्रकार निर्धारित करें तथा किस प्रकार उपाय निकालें, परिशुद्धता तथा सुसंगति की जांच करे, संचालन समिति के कार्य इत्यादि पर बीबीएस पर्यवेक्षकों तथा संचालन समिति के सदस्यों का कार्यस्थलीय व्यावहारिक प्रशिक्षण का आयोजन करें।

நிறுவனங்களில் நடத்தை சார்ந்த பாதுகாப்பு (BBS) அமைப்பு

நடத்தை செய்பாடுகள் சார்ந்த பாதுகாப்பு (பிபிஎஸ்) என்பது நிறுவன ஊழியர்களுக்கான பாதுகாப்பு கலாச்சாரம். இதன் முதல் பயனாளி நிறுவனத்தின் பாதுகாப்பு துறை. இதன் மூலம் பல நடத்தை அடிப்படையிலான பாதுகாப்பு பயிற்சி பெற்ற பயிற்சியாளர்கள் இந்த நடத்தை சார்ந்த பாதுகாப்பு முறையை நாடு முழுவதும் உள்ள நிறுவனங்களில் பணியாற்றும் ஊழியர்களிடம் கொண்டு சேர்க்கிறார்கள்.

ஒரு நிறுவனத்தின் பிரதான குறிக்கோள் பூஜ்ய நிலை விபத்து புள்ளியை அடைவதாகும். நிறுவனத்திற்கு 100% ஊழியர்களின் பங்களிப்பு நிறுவனத்தின் பாதுகாப்பிற்கு மிகவும் அவசியமானது. இது தான் நடத்தை சார்ந்த பாதுகாப்பு முறை ஆகும்.

இதை அடைய அனைத்து நிறுவன ஊழியர்களும் தயார் படுத்தப்பட வேண்டும். இதை ஊழியர்களுக்கு அறிமுகப்படுத்தும் முன்னர் துறை தலைவர்களும் மேலாளர்களும் நடத்தை சார்ந்த பாதுகாப்பை பற்றி உணர வேண்டும்.

நடத்தை சார்ந்த பாதுகாப்பு அமுல் படுத்தப்படும் முன்னர், தொழிற்சாலை பாதுகாப்பு என்பது காவல்துறை ஆய்வைப்போல செயல்பட்டு வந்தது. நடந்தை சார்ந்த பாதுகாப்பு கருத்துப்பரிமாற்றத்தை அதிகாரிகளிடம் மேம்படுத்தியதோடு இல்லாமல் ஊழியர்களிடமும் அவர்களது. பாதுகாப்பு மனப்பான்மையை அதிகப்படுத்தியது. ஊழியர்கள் தங்களது வேலையில் அதிக ஆர்வம் காட்டினர். ஊழியர்கள் தங்களது செயல்முறைகளை அறிந்து வைத்திருப்பதால், நடத்தை சார்ந்த பாதுகாப்பு முறையை நன்றாக அமுல் படுத்த முடியும்.

நடத்தை சார்ந்த பாதுகாப்பு (BBS) அமைப்பின் தேவை

ஆராய்ச்சியும், அனுபவமும் கீழ்க்கண்டனவற்றை சுட்டிக் காட்டுகின்றன.

- 90% மற்றும் அதற்கு அதிகமான விபத்துக்கள் பாதுகாப்பற்ற மனித செயல்களால் நிகழ்கின்றன.
- 50% விபத்துக்கள் பாதுகாப்பற்ற நடவடிக்கைகளால் நிகழ்கின்றன. இவைகளை எந்த தொழிற்சாலைகளிலும் எந்த நிலையிலும் கண்டறியலாம்.
- 20% 30% விபத்துக்கள் ஊழியர்களிடம் பாதுகாப்பு குறித்த அறியாண்மை மற்றும் அதனால் ஏற்படும் பாதுகாப்பற்ற முறைகளால் ஏற்படுகின்றன.

- இந்த பாதுகாப்பற்ற முறைகள் விபத்துகள், காயங்கள் மற்றும் விபத்தின் அருகாமையின் தப்பித்தல் போன்றவைகளுக்கு மையாமாகும். பாதுகாப்பான முறைகளை பின்பற்றினால் இவற்றை தவிர்க்கலாம்.
- இதனால் நம்முடைய லட்சியமும், செயலும் பாதுகாப்பான முறைகளை பின்பற்ற செய்ய வேண்டியதாக இருத்தல் வேண்டும்.

நடத்தை சார்ந்த பாதுகாப்பு முறையின் (BBS) வெற்றி என்பது அனைத்து ஊழியர்களுக்கும் சக்தியளித்து அவர்களே பொறுப்பேற்று பாதுகாப்பு வழிகளை பின்பற்ற செய்வதாகும்.

நடத்தை சார்ந்த பாதுகாப்பு (BBS) முறை, என்பது OHSAS– ன் இணக்கம் OHSAS 18001 : 2007 ஆவணம் OH&S மேலாண்மை முறைகளை கொண்டுள்ளது.

(a) உட்பிரிவு 4.4.3.2 (பங்கெடுத்தல் மற்றும் கருத்து பரிமாற்றம்): ஒரு நிறுவனம் தனது வழிமுறைகளை அமல்படுத்த வேறூன்ற, பராமரிக்க. தனது ஊழியர்களை ஆபத்துக்களை கணிக்கவும், கடினமான பணிகளை கண்டறியவும், எல்லைக்கோடுகளை வரையறுக்கவும் உட்படுத்தலாம்.

(b) OHSAS 18001 : 2007 பிரிவு 4.3.1 (ஆபத்துகளை கணிக்க, கடினவேலைகளை கண்டறிவது மற்றும் எல்லைக்கோடுகள் வரையறுத்தல்) படி ஒரு நிறுவனம் ஆபத்துகளை கணிக்கவும், கடின வேலைகளை கண்டறியவும், மனித நடத்தை சார்ந்த அவர்களின் திறமை மற்றும் இதர தன்மைகளை ஆராய கண்டறியவும் நெறிமுறைகளை உருவாக்கலாம்.

(c)OHSAS 18001 : 2007, பிரிவு 4.4.2 ன் படி (திறமை, பயிற்சி) மற்றும் விவரங்களை தெரிந்திருத்தல்) ஒரு நிறுவனம் தன்கீழ் பணியாற்றும் ஊழியரிடம் OH&S ன் பின் விளைவுகள், ஆற்றல், வேலை நடவடிக்கைகள், நடத்தை மற்றும் முதன்மையான செயல்பாடு ஆகியவற்றை அறிய நடவடிக்கை எடுக்கலாம்.

நடத்தை பாதுகாப்புமுறை OHSAS 18001 : 2007 – யை தன் பயிற்சியில் உள்ளடக்கி உள்ளது. அதனால் நடத்தை பாதுகாப்பு முறை ஊழியர்களின் பங்களிப்பு மற்றும் Involvement அதிகப்படுத்தி அவர்களை ஆபத்துகளை கணிக்கவும், கடின வேலைகளை கண்டறியவும், எல்லைக் கோடுகளை ஆராயவும் மற்றும் பாதுகாப்பான முறைகளை பின்பற்றவும் உதவி செய்கிறது. மேலும் இந்த நடத்தை பாதுகாப்பு முறை ஊழியர்களின் நடவடிக்கைகளை அவர்களின் பாதுகாப்பான மற்றும் பாதுகாப்பற்ற வழிமுறைகளின் மூலம் மாதா மாதம் கணக்கெடுக்க வழிமுறை செய்கிறது. நடத்தை சார்ந்த பாதுகாப்பு முறை மேலே சொல்லியுள்ள மூன்று வகைகளை எளிதாக நடைமுறைபடுத்த வகைசெய்கிறது.

நடத்தை பாதுகாப்பு சார்ந்த (BBS) பயிற்சி நோக்கம்

இப் பயிற்சி முகாமின் மூலம் அனுபவம், பாதுகாப்பான வழிமுறைகளை வெளிக்கொணர்ந்து கவனித்தல் மற்றும் கருத்து பரிமாற்றம், நடத்தை பாதுகாப்பு முறை அமுல்படுத்துதல் போன்றவற்றை கற்றுக் கொடுக்கிறது.

பயிற்சி முகாம் ஆழ்ந்த கருத்துக்களை புரிந்து கொள்ளவும், நடத்தை பாதுகாப்பு முறையை பயன்படுத்தவும், பாதுகப்பற்ற முறைகளை களையவும், விபத்துகளை தடுக்கவும், பாதுகாப்பு முறைகளை மேம்படுத்தவும், நிறுவனங்களுக்கு உதவுகின்றது.

பயிற்சி முகாமின் பொதிவு

நடத்தை பாதுகாப்பு முறைகளை அமுல்படுத்தும் முன்னர் பயிற்சி அவசியமானதாகும். அதனால் பயிற்சி முகாமில் கீழ்கண்டவைகள் சேர்க்கப்பட்டுள்ளன.

- நடத்தை பாதுகாப்பு எண்ணங்கள்
- நடத்தை பாதுகாப்பு முறை செயல்முறை
- பாதுகாப்பான & மற்றும் பாதுகாப்பற்ற நடத்தைகளின் அளவீடு
- பாதுகாப்பான மற்றும் பாதுகாப்பற்ற பதிவுகள் மற்றும் Display
- நடத்தை பாதுகாப்பு முறையின் அமலாக்கம்
- செயல்முறை திட்டம் மற்றும் இதன் செயலாக்கமுறை

நடத்தை பாதுகாப்பு முறையின் (BBS) செயல்பாடு

நடத்தை பாதுகாப்பு முறையானது சிறப்பான மற்றும் நவீனமான முறையாகும். நடத்தை முறை பாதுகாப்பு செயல்முறை என்பது ஊழியரின் நடவடிக்கைகளை கண்காணித்து அவர் தன்னுடைய நடத்தை முறைகளை 100% பாதுகாப்பானதாக மாற்றவைக்கும். மேலும் அவரை தன்னைத்தானே அறிந்து கொள்ளவும், மற்றவர்களின் நடவடிக்கைகளை கண்காணிக்கவும் அவர்களை பாதுகாப்பாக நடக்க வைத்துக் கொள்ள உதவும்.

எட்டுவழிகள் இந்த நடத்தை முறை பாதுகாப்பு முறையை வழிவகை செய்கின்றன.

1. தனிமனித காப்பு உபகரணங்கள் (PPE)
2. சுற்றுப்புறத்தூய்மை
3. உபகரணங்கள் உபயோகித்தல்
4. வேலையின் போது உடலை சரியான இடத்தில் நிறுத்துதல்.
5. பாகங்களை கையாளுதல்
6. தகவல் பரிமாற்றம்

7. நடமுறைகளை பின்பற்றுவது
8. கவனமாக செய்தல்

ஒரு சரிபாக்கும் பட்டியல் மூலம், தேர்வடைந்து நடத்தை பாதுகாப்பு முறை கண்காணிப்பாளர்கள் தகவல்களை மேல் கூறப்படும் 8 நடத்தைகளின் மூலம் கண்காணித்து அவர்களை நடைமுறை படுத்துபவர்களின் ஒரு நடைமுறையில் தினமும் ஒருவரை கண்காணிப்பர். இது தினமும் நடைபெறுவதால். பாதுகாப்பு கலாச்சாரத்தை வித்திட வழி வகுக்கும். இதன் மூலம் பாதுகாப்பான மற்றும் பாதுகாப்பற்ற நடைமுறைகளை தினமும் கண்காணித்து களைய முடியும்.

நடத்தை பாதுகாப்பு முறை ஒருவரின் பாதுகாப்பற்ற முறைக்கு அவரை தண்டிக்காமல் அவரை மேலும் மேலும் பாதுகாப்பான முறையை பின்பற்றுவதற்கு அவரை அறிவுறுத்தும். ஒருவர் பாதுகாப்பான முறையை பின்பற்றினால் அவரை மேலும் ஊக்கப்படுத்தும். பாதுகாப்பான மற்றும் பாதுகாப்பற்ற முறைகள் கண்காணிக்கப்பட்டு அவைகளை காட்சிபடுத்தப்படும். . மேலும் எந்த பாதுகாப்பற்ற சூழ்நிலையினால், பாதுகாப்பற்ற நடைமுறைகள் உருவாகிறது என்பது பற்றி விவாதிக்கப்படும்.

நடத்தை பாதுகாப்புமுறை என்பது தகவல்களை வைத்து முடிவெடுக்கும் முறையாகும். இதனால் ஒவ்வொரு ஊழியரிடமும் வித்தியாசத்தைக் கண்டு அவர்களின் பாதுகாப்பு தனித்திறமையை கண்டறிய முடியும். ஊழியர்கள் தங்களை சுற்றி நடக்கும் விஷயங்களை தெரிந்து தங்களை சரிப்படுத்திக் கொள்ளும் திறமை மிக்கவர்கள். நடத்தை பாதுகாப்பு முறை எல்லா துறைகளைப் பற்றியும் விவரம் அளிக்கும் இது ஒரு கூட்டு முயற்சி. இது மக்களால் நிறுவனங்களில் நடத்தப்படும். நடத்தை பாதுகாப்பு முறை என்பது பாதுகாப்பான முறைகளை கையாள, கடினமான வேலைகளை பாதுகாப்பான செயல்முறைகளை கண்டறியவே ஒழிய, பாதுகாப்பு சட்டங்களை ஊழியர்கள் மீது திணிப்பதற்கில்லை. ஒரு குழு இந்த முறைகளை கண்காணித்து நடைமுறைப் படுத்தும்.

நடத்தை பாதுகாப்பு முறை என்பது நிர்வாகத்தின் கருவி என்பதை விட ஊழியர்களின் பாதுகாப்பான வாழ்வு முறைக்கு நிர்வாகத்தின் உதவி என்று கூறுவது சாலச்சிறந்ததாகும்.

நடத்தை சார்ந்த பாதுகாாப்பு முறை (BBS) முக்கோணம்

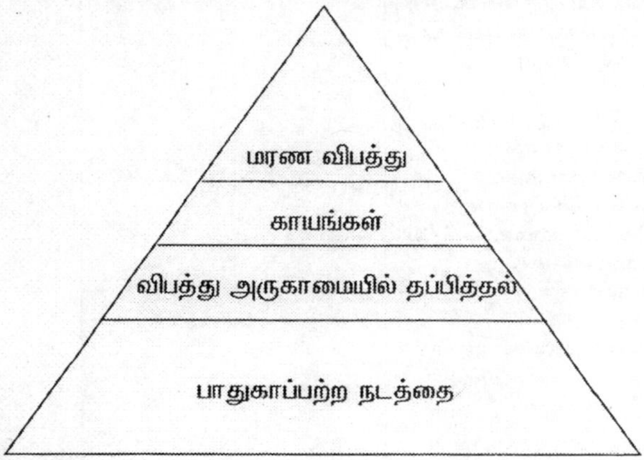

பயிற்சியாளர் பற்றிய குறிப்பு:

டாக்டர் எச்.எல். கைலா அவர்கள் மும்பையில் சைக்காலஜி துறை பேராசிரியர். இத்துறையில் 25 வருடத்திற்கும் மேற்பட்ட அனுபவமிக்கவர். ஐசிஎஸ்எஸ்ஆர், யூஜிசி, என்எஸ்சி மற்றும் உலக சுகாதார நிறுவனம் ஆகியவற்றின் பல திட்டங்களில் இணைந்து பணிபுரிந்தவர்.

இத்துறையில் 12 புத்தகங்களை எழுதியுள்ளார். செய்தித்தாள் மற்றும் முக்கியமான பத்திகைகளுக்கு அதிகமான கட்டுரைகள் எழுதியுள்ளார்.

தொலைக்காட்சியில் இத்துறை சம்பந்தமாக நிறைய செய்திகளை கூறியுள்ளார்.

பல வெளிநாடுகளில் நடக்கும் சைக்கலாஜி துறை சம்பந்தமான கருத்தரங்குகளில் இந்தியாவிற்காக கலந்து கொண்டுள்ளார்.

ஜர்னல் ஆப் சைக்காலஜி பத்திரிகையின் எடிட்டராகவும், கவுன்சிலர்ஸ் அசோசியேஷன் ஆப் இந்தியாவின் உதவி தலைவராகவும் உள்ளார்.

இந்தியாவின் தொழில் சம்பந்தமான சைகாலஜி துறையில் சிறந்த BPS பயிற்சியாளராவார்.

சரிபார்க்கும் பட்டியல்

	நடத்தை முறை	பாதுகாப்பாக உள்ளது	பாதுகாப்பாக இல்லை	மேம்படுத்துவதற்கான கருத்துகள்
1.	தனிநபர் காப்பு உபகரணங்கள்– கண்ணாடி, கையுறை, ஹெல்மெட் போன்றவை உள்ளனவா.			
2.	தொழில் செய்யும் இடம் குப்பை, தேவையற்ற சாமான்கள், நடைபாதை அடைப்புகள் இல்லாமல் உள்ளனவா என்றும் உபகரணங்கள் சரியான இடத்தில் உள்ளனவா என்பதை சோதிக்கவும்.			
3.	செய்யவேண்டிய வேலைக்கு சரியான உபகரணங்கள் நல்ல நிலையில் உள்ளனவா.			
4.	ஊழியர் தன்னை சரியான இடத்தில் நிலைப்படுத்தி உள்ளாரா?			
5.	பொருட்களை காக்கும் போதும், தள்ளும்போதும், சரியான உபகரணங்கள் உபயோகிக்கப் படுகின்றனவா?			
6.	வாய்மொழி மற்றும் வாய்மொழியற்ற தகவல் பரிமாற்றங்கள் சரியான முறையில் உள்ளனவா?			
7.	பாதுகாப்பு வழிமுறைகள் சரியான முறையில் பின்பற்றபடுகின்றனவா?			
8.	வேலையில் கவனமாக இருக்கிறார்களா?			
9.	கைபேசி பயன்படுத்துகிறார்களா?			

1. જ્યારે આપ કોઈપણ અસુરક્ષિત વર્તણૂક જુઓ, તો તેને તાત્કાલિક સ્થળ પર જ સુધારો.
2. BBS એટલે ભાઈભાઈની સુરક્ષા
3. અસુરક્ષિત વર્તણૂક દરેક નીયરમીસ, ઈજા અથવા અકસ્માતમૃત્યુનું મૂળભૂત કારણ હોય છે.
4. BBS માં આપણું ધ્યેય "શૂન્ય અસુરક્ષિત વર્તણૂક" છે, નહીં કે "શૂન્ય અકસ્માત"
5. જ્યારે તમે કોઈપણ નીયરમીસ રિપોર્ટ કરો છો, ત્યારે તે દરમિયાન થતી અસુરક્ષિત વર્તણૂક રિપોર્ટ કરો.
6. અસુરક્ષિત વર્તણૂક કોઈપણ કરી શકે છે, જેમ કે કામદાર, સુપરવાઈઝર, એન્જિનીયર, વાઈસ પ્રેસીડન્ટ તથા ઉંમર, અનુભવ કે પદનું કોઈ મહત્વ નથી.
7. અસુરક્ષિત વર્તણૂક એ સ્વભાવિક અને નિસ્વાર્થ હોઈ શકે. (તમે મને બચાવો, હું તમને બચાવીશ)
8. BBS એ કોઈના પર ગુસ્સો, કોઈનું અપમાન અથવા કોઈના પર આરોપ કે ભય પેદા કરવા માટે નથી.
9. BBS વ્યક્તિને રૂબરૂ મળી સુરક્ષિત વર્તણૂક માટેની વિનંતી તથા તેની સુરક્ષિત વર્તણૂક માટે આભાર વ્યક્ત કરવામાં માને છે.
10. BBS કામ અટકાવવા માટે નથી પરંતુ અસુરક્ષિત વર્તણૂક અટકાવવા માટે છે.
11. સેફ્ટી થકી ઉત્પાદન અટકે અથવા મોડું પડે છે, એ વાત મિથ્યા છે.
12. અસુરક્ષિત વર્તણૂક ઉત્પાદનમાં ઘટાડો તથા વખતોવખત મળતા પ્રોત્સાહક ઈનામોની રકમમાં ઘટાડા તરફ દોરે છે.
13. BBS ભાઈચારો, સેફ્ટી, હેલ્થ તથા ભાઈની ઉત્પાદનમાં વધારો કરે છે.
14. BBS એ પાયાથી ઉપર તરફ જતો સિધ્ધાંત છે, તે સકારાત્મક અને લોકો દ્વારા ચાલતો લોકોની સુરક્ષા માટેનો કાર્યક્રમ છે.
15. BBS એ સામાન્ય બુધ્ધિની વાત નથી તે માનવીય વર્તણૂકનું સિધ્ધ થયેલ વિજ્ઞાન છે.

16. માનવીય વર્તણૂક નીચે જણાવેલ નવ પ્રકારે નોંધી શકાય.
    - વ્યક્તિગત સુરક્ષાના ઉપકરણ
    - કાર્યક્ષેત્રમાં ગૃહવ્યવસ્થા
    - સાધનો તેમજ સામનોનાં ઉપયોગ સંબંધી
    - કાર્યસ્થળ પર વ્યક્તિના શરીરની સ્થિતિ અથવા સુરક્ષા
    - માલસામાનની હેરફેરી માલસામાનની હેરફેરી
    - સંદેશા વ્યવહાર
    - કાર્યપધ્ધતિનું અમલીકરણ
    - દ્રશ્ય પ્રણાલિ
    - કાર્યસ્થળ પર મોબાઈલ ફોનનો ઉપયોગ
17. કોઈપણ અસુરક્ષિત વર્તણૂકનું મૂલ્ય તમારા કે બીજા માટે પણ કદી ઓછું આંકશો નહીં.
18. BBS આપણને ઘણા મોટા કુટુંબનો સભ્ય બનાવે છે.
19. સેફ્ટીની શરૂઆત વ્યક્તિગત સુરક્ષાનાં સાધનોથી થાય છે, અંત પણ વ્યક્તિગત સુરક્ષાનાં સાધનોથી હોય, પરંતુ એ સ્ટાન્ડર્ડ ઓપરેટીંગ પ્રોસીજર સુધી જાય છે.
20. BBS કોઈપણ અસુરક્ષિત વર્તણૂક અવગણવા માટે નથી.
21. BBS આપણી વ્યકિગત છાપ પર સેફ્ટીનો દ્રષ્ટિકોણ ઉમેરે છે.
22. BBS અવલોકનકર્તા એ ગરૂડ સમાન છે.
23. BBS અવલોકનકર્તા એ કોમ્પ્યુટરમાં વાઈરસ સામે લડનાર સોફ્ટવેર જેવા છે.

24. એકબીજાની સેફ્ટીના પહેરેદાર બનીને આપણે HES અને HZMC ની હદની રક્ષા કરી શકીયે છીએ.
25. BBS એ હાલની ચાલતી કે કોઈ બીજી સેફ્ટીની પ્રણાલિ ની જગ્યા લેવા માટે નથી. પરંતુ તેને વધુ સબળ અને પ્રભાવશાળી બનાવવા માટે છે. કારણ કે એમાં દરેક એકબીજાની સંભાળ લે છે.
26. BBS થી સેફ્ટીને લાઈન ફંકશન એટલે કે સેફ્ટી વિષે કાર્યસ્થળ પરના સભ્યોની સીધી જવાબદારી સમજવામાં મદદ મળશે.
27. BBS ના અમલીકરણથી OHSAS 18001 ની જરૂરિયાત પણ સંતોષાશે.
28. BBS ના અમલીકરણથી જ્યારે તમે બીજી વ્યક્તિને મળશો ત્યારે તે તમારા સકારાત્મક વલણથી અંજાય જશે.
29. BBS એ ફરજિયાત નથી.પરંતુ એ એક લગાવ છે.
30. BBS એ લોકોની તકલફિ વિષે વિચારે છે.
31. BBS એ માનવીય સુરક્ષા પૂરી પાડે છે, જેમાં કામની સુરક્ષા પણ આવી જાય.
32. બીજાનું જીવન બચાવવું એ આપણી એક સ્વભાવિક વર્તણૂક છે, તેથી BBS ના અમલીકરણમાં મને સરળતા રહેશે.
33. જ્યારે આપણા કાર્યક્ષેત્રમાં હજારોની સંખ્યામાં અસુરક્ષિત વર્તણૂક જોવા મળે છે ત્યારે ખરેખર આપણને BBS નામની દવાની જરૂર છે.
34. અવલોકનકર્તા પાસે બે ખાસિયત હોવી જરૂરી છે.
    - એ જેનું અવલોકન કરે છે. તે વ્યક્તિને વાત / રજૂઆત કરવા દે.
    - જ્યારે તે છૂટો પડે ત્યારે, જેનું અવલોકન કરે છે તેના મોં પર સ્મીત જુએ.
35. BBS એ સુરક્ષા માટેની પરિક્રમા છે.

36. BBS એ કંઈ નામ કમાવવાની, ગુનો પ્રસ્થાપિત કરવાની કે સજા કરવાની પ્રણાલિ નથી, એ તો સુરક્ષિત વર્તણૂકોનું ખાતું છે.
37. BBS એ જીવન જીવવાની કળા છે.
38. BBS એ ડો. ભૂપેન હજારિકાના મતે માણસોની એકબીજા માટેની સુરક્ષિત પ્રણાલિ છે. (મનાવે મનાહર બાબે)
39. BBS એ અસુરક્ષિત વર્તણૂકને સમયસર સુધારવાની પ્રણાલિ છે.
40. BBS શૂન્ય અસુરક્ષિત વર્તણૂક અથવા શૂન્ય અકસ્માત માટેનું વાતાવરણ સર્જે છે.
41. BBS ના અમલીકરણ દરમિયાન અસુરક્ષિત વર્તણૂક સામેની લડતમાં યુધ્ધવિરામ આવતો નથી.
42. BBS એ તાત્કાલિક જીવન બચાવવાની કાર્યપ્રણાલિ છે અને તે પરિણામલક્ષી છે.
43. BBS ની સફળતા માટે જરૂરી છે.
    - તાલીમ પામેલ અવલોકનકર્તા દ્વારા સતત અવલોકન.
    - BBS કાર્યપ્રણાલિમાં સતત જોડાયેલા રહેવું અને અસરકારતા માપતા રહેવું.
44. BBS એ મતભેદ ઊભા નથી કરતું પરંતુ મતભેદ ઘટાડે તેવી પ્રણાલિ છે.
45. BBS ની મદદથી અસુરક્ષિત વર્તણૂક અને સુરક્ષિત પરિસ્થિતિ વચ્ચેનો સંબંધ સરળતાથી પ્રસ્થાપિત કરી શકાય છે.
46. હવે હું સમજ્યો છું કે વર્તણૂક કેવી હોવી જોઈએ, કેવી રીતે માપી શકાય, કેવી રીતે સુધારી શકાય અને BBS નું સાચુ અવલોકન કેવી રીતે કરી શકાય.

Index